T0137080

# Privacy-Preserving in Mobile Crowdsensing

Chuan Zhang • Tong Wu • Youqi Li • Liehuang Zhu

# Privacy-Preserving in Mobile Crowdsensing

 Springer

Chuan Zhang
Beijing Institute of Technology
Beijing, China

Tong Wu
Beijing Institute of Technology
Beijing, China

Youqi Li
Beijing Institute of Technology
Beijing, China

Liehuang Zhu
Beijing Institute of Technology
Beijing, China

ISBN 978-981-19-8317-7      ISBN 978-981-19-8315-3    (eBook)
https://doi.org/10.1007/978-981-19-8315-3

This Springer imprint is published by the registered company Springer Nature Singapore Pte Ltd.
The registered company address is: 152 Beach Road, #21-01/04 Gateway East, Singapore 189721,
Singapore

*We pay our highest respect and gratitude to the editors for their support in publishing this book and thank all reviewers and peers for their precious advice in bettering this book. We also thank the School of Cyberspace Science and Technology in Beijing Institute of Technology and all our colleagues in the lab for their support and help in completing this book. We further appreciate our families for their unconditional love. We still remember the holidays and vacations when we were finishing this book while not keeping our families company. Dr. Zhang dedicates this work to his parents, father Wuping Zhang and mother Xinfeng Li, who have brought him up and sacrificed so much, as well as his wife, Jinyang Dong. He could never have done this without his parents' and wife's love, support, and constant encouragement. A sincere appreciation to all family members for their continuous love. Dr. Zhang gives his gratitude to his supervisor, Liehuang Zhu. Dr. Wu would like to thank her father, Shuzhong Wu, and her mother, Weidong Lu, for their endless love and care. She gives her gratitude to her postdoctoral supervisor, Liehuang* vi

*Zhu. Finally, she is grateful for everything she has experienced, acquired, and expected in the Beijing Institute of Technology. Dr. Li would like to thank his mother, Huanv Fu, for her continuous love and support for his life. He gives his gratitude to his postdoctoral supervisor, Liehuang Zhu. Finally, and particularly, he gives a profound thanks to the co-authors and our students, in continuing appreciation of their support and many other contributions. Dr. Zhu would like to thank his wife, Fang Guan; daughter, Jie Zhu; mother, Lingying Zeng; brother, Liehui Zhu; and many other relatives for their continuous love, support, trust, and encouragement throughout his life. Without them, none of this would have happened.*

# Foreword

This book focuses on picturing mobile crowdsensing from a few perspectives, including architecture, key technologies, security principle, privacy requirements, service models and system framework, etc. The main content of this book is derived from the most advanced technical achievements or breakthroughs in the field. A number of representative mobile crowdsensing services will be covered by this book, such as environmental sensing, traffic supervision, and health data collection. Both theoretical and practical contents will be involved in this book in order to assist the readers to have a comprehensive understanding of the mechanism on achieving user privacy protection during data collection and data processing in mobile crowdsensing.

This book facilitates students and learners to acquire some basic knowledge of task assignment and truth discovery in mobile crowdsensing applications and supports both practitioners and scholars to find some new insights in mobile crowdsensing.

Beijing, China

Chuan Zhang
Tong Wu
Youqi Li
Liehuang Zhu

# Preface

Mobile crowdsensing (MCS) is a novel data sensing paradigm that integrates mobile sensing and crowdsourcing. It is centered on the users with various kinds of sensors (e.g., pressure meter, accelerometer, gyroscope, etc.), and it can complete complex and large-scale social sensing tasks, including environmental sensing, traffic monitoring, difficult problem solving, etc., through human-sensor cooperation and human-human cooperation. Toward the enhanced secure and privacy-preserving MCS, this book introduces a few significant aspects of MCS, including the fundamental knowledge and technologies of MCS, privacy-preserving task allocation, privacy-preserving truth discovery, and future research trends.

MCS has the advantages of high sensing efficiency, long life cycle, and low deployment cost, which makes better use of the wisdom and resources of the crowd and achieves a wide range of sensing coverage and high efficiency of data acquisition. Users with mobile devices participate in the MCS systems as the basic sensing units and gain rewards by providing the data. On the other hand, it is crucial to consider the risk of privacy leakage, since the data submitted by the users may contain private personal information such as identity, gender, location, hobbies, etc. The research works in this book will help to improve the capability of security and privacy in MCS applications, and provide critical thinking and solid references for MCS. This book has certain academic and applicable value on privacy preservation in MCS.

This book focuses on picturing privacy-preserving in MCS techniques from a few perspectives, which are architecture, models, key technologies, security principle, privacy requirements, system model, framework, etc. The main contents of this book derive from the most updated technical achievements or breakthroughs in the field. Both theoretical and practical contents will be involved in this book in order to assist the readers to have a comprehensive understanding of privacy-preserving in MCS.

This book facilitates students and learners to acquire basic knowledge of MCS and supports practitioners and researchers in finding new insights into privacy-preserving approaches in MCS.

Beijing, China                                                                           Chuan Zhang
                                                                                                 Tong Wu
                                                                                                Youqi Li
                                                                                        Liehuang Zhu

# Acknowledgements

The authors of this book would like to acknowledge our editors for their help and guidance, and the students at Beijing Institute of Technology, Haotian Deng, Mingyang Zhao, Chenfei Hu, Weijie Wang, and Xiaochen Ma, for their contributions to the accomplishment of this book. All the supports, advice, and encouragement from peer experts, scholars, students, and the team members at the Beijing Institute of Technology are remarkably meaningful and invaluable for the accomplishment of this book. The authors sincerely appreciate all the individuals and organizations who reinforce and raise the quality of this book.

Beijing, China
August 2022

Chuan Zhang
Tong Wu
Youqi Li
Liehuang Zhu

# Contents

# Acronyms

| | |
|---|---|
| COA | Ciphertext-Only Attack |
| CP | Crowdsensing Platform |
| CPA | Chosen-Plaintext Attack |
| CSP | Cloud Service Provider |
| DR | Data Requester |
| GPTA | Geometric Range Query Based Privacy-Preserving Task Allocation |
| IND-CPA | Indistinguishability Under Chosen-Plaintext Attack |
| kNN | k-Nearest Neighbor |
| KPA | Known-Plaintext Attack |
| LPTD | Lightweight Privacy-Preserving Truth Discovery |
| MCS | Mobile Crowdsensing |
| PCDD | Public-Key Cryptosystem Supporting Distributed Decryption |
| PPTA | Privacy-Preserving Task Allocation |
| RPTD | Reliable and Privacy-Preserving Truth Discovery |
| SATE | Privacy-PreServing Non-InterActive Truth DiscovEry |
| SE | Searchable Encryption |
| SHVE | Symmetric Hidden Vector Encryption |
| SU | Sensing User |
| TA | Trusted Authority |
| TD | Truth Discovery |
| TPE | Threshold Predicate Encryption |

# Part I
# Overview and Basic Concept of Mobile Crowdsensing Technology

# Chapter 1
# Introduction

**Abstract** Mobile crowdsensing (MCS) is a new data sensing model that combines mobile sensing and the idea of crowdsourcing. It is centered on users with multiple sensor resources and accomplishes complex, large-scale social sensing tasks through human-object collaboration and human-human collaboration. As the user is the subject of MCS, the sensory data collected and contributed by the user may contain sensitive information related to his/her identity, gender, location, hobbies, etc., exposing the user to the risk of privacy leakage. This chapter first introduces the background of MCS and then describes typical MCS models. Existing privacy-preserving task allocation and truth discovery schemes are then briefly reviewed. Finally, the organization of this book is presented.

## 1.1  Background

The rapid development of the Internet of things (IoT) has brought profound changes to human life and industry production. By deploying devices such as ultrasonic sensors and loop coil sensors on roads, traffic management departments can effectively monitor traffic conditions at different locations; by placing devices such as blood pressure sensors and pulse sensors on patients, medical institutions can observe the physical condition of the patients in real time; by deploying devices such as dust and sulfur dioxide sensors near factories, environmental regulatory department can be informed of the air quality around the factory. Thanks to the increasing abundance and popularity of various sensors (e.g., microphones, accelerometers, manometers, gyroscopes), the IoT is also broadening the scope of relevant services, influencing and promoting the development of smart homes, smart factories, smart cities, etc. Data plays a vital role in IoT, and the applications in IoT are closely related to the scope and type of the sensing data from various devices. However, data collection in the IoT becomes increasingly difficult as the demand for applications continues to increase, the scale of services gradually expands, and the cost of sensor deployment and maintenance continues to rise. For example, building smart cities requires the deployment of a large number of diverse sensing sensors across major city facilities, but the huge deployment and maintenance overheads and complex sensor management issues pose a significant challenge to the development of smart cities. Due to the above reasons, MCS has emerged.

MCS is an IoT data sensing model combining mobile sensing and crowdsourcing ideas. Unlike sensor-based data sensing, MCS introduces people into the data sensing environment, making full use of people's external resources (e.g., mobile devices) and internal resources (e.g., background knowledge, education level, and behavioral habits), transforming conventional object-object data sensing into human-object and human-human collaborative data sensing, greatly expanding the scope and scale of data sensing, and significantly reducing the cost of data collection. Ganti first proposed the concept of MCS in 2011 [1]. Then, Guo et al. further enriched the definition of MCS [2] and formalized the framework of MCS [3]. Later, Liu et al. summarized the common techniques used in mobile crowdsensing applications [4], aiming to reduce the consumption of resources such as computation and communication in MCS, while Capponi et al. [5] summarized the current main applications of MCS and the challenges it faces and how to address them. Finally, Liu et al. [6] developed a mobile crowdsensing system platform (Crowd OS) to help city authorities with urban refinement.

MCS has attracted a great deal of attention from academia and industry because of its capability on effectively using the wisdom of groups and society's idle resources. Many scholars have proposed MCS-based application for different scenarios in IoT. For example, in traffic management, Tang et al. [7] designed a lane information collection scheme based on MCS using in-vehicle GPS to help relevant authorities to get the current traffic information; Teng et al. [8] proposed an indoor-outdoor navigation scheme based on MCS to help users quickly find their way from an indoor (e.g., underground station) to a specific outdoor destination; and Ni et al. [9] designed a navigation service based on MCS using devices such as mobile phones or in-vehicle sensors. In terms of environmental awareness, Zappatore et al. [10] designed a noise data collection scheme based on MCS; Pan et al. [11] proposed an air quality detection scheme based on MCS using a camera-equipped mobile device. In recent years, due to the continuous development of cloud computing, edge computing, 5G/6G communication technologies, and the widespread use of smart terminals equipped with various sensors (e.g., smartphones, smart tablets, smart bracelets), many commercial applications based on MCS have emerged. Examples include WAZE and Google Maps for traffic navigation; Zhihu and Baidu for consulting service; Yelp and Meetup for service recommendations; Collision Awareness for road information; and Be My Eyes for life support for visually impaired users.

MCS is a human-centered data sensing model, which leads to data in MCS often containing sensitive information related to people for human involvement [12–17]. From the perspective of the life cycle of data in MCS, MCS can be roughly divided into the data collection phase and data processing phase. In the data collection phase, in order to improve the accuracy and efficiency of data collection, sensing users usually send their task interests to the mobile crowdsensing platform, to find a suitable sensing user based on the task requirements of the data requester (e.g., task content requirements, geometrical range requirements), known as "task allocation" [18].

However, as task information is usually closely related to the user's spatiotemporal information, such as time, location, and trajectory, internal/external attackers may infer the user's gender, address, identity, hobbies, and education level from the user's task information. Once the data is uploaded to the mobile crowdsensing platform, the mobile crowdsensing platform needs to provide the results back to the data requester. However, the quality of the data submitted by the user often varies due to their different device quality, knowledge background, and interests. Therefore, it is necessary to get the real sensory data from the data submitted by users, called "truth discovery" [19]. In the truth discovery process, sensing data is directly exposed to the mobile crowdsensing platform. On one hand, the attackers may infer the user's identity, location, religious beliefs, behavioral habits, and other private information based on the user's sensing data. On the other hand, the attackers could resell the user's data to other companies or individuals, which violates the user's privacy and economic interests. In fact, as a third-party platform, the mobile crowdsensing platform is usually not fully trusted. In recent years, there are numerous incidents of third-party platforms stealing or illegally using users' information. For example, in March 2018, Cambridge Analytica was exposed for improper use of tens of millions of Facebook users' data;[1] in August 2018, approximately 500 million pieces of user data were suspected to be compromised and sold by the Huazhu Group;[2] in August 2020, applications such as NetEase Open Class and EasyChange were disclosed by the Ministry of Industry and Information Technology for illegally collecting or using users' private information;[3] in August 2020, Facebook in the United States was suspected of illegally collecting user biometric data through Instagram app without user consent;[4] and in November 2020, Yuantong was interviewed by the Shanghai Internet Information Office for leaking 400,000 pieces of personal privacy information by an "insider".[5] Therefore, how to protect user privacy during data collection and data processing is an important research content of MCS.

From the perspective of privacy protection in MCS, this book investigates efficient and privacy-preserving task allocation schemes and truth discovery schemes to address the current problems of MCS in data collection and data processing such as the risk of privacy leakage, low efficiency, and single function. At the application level, the proposed privacy-preserving content-based task allocation scheme helps data requesters find sensing users who satisfy their task content requirements; the proposed privacy-preserving location-based task allocation scheme supporting arbitrary geometric range queries helps data requesters find sensing users that satisfy their geometric range needs; the proposed privacy-preserving truth discovery scheme with truth transparency helps the mobile crowdsensing platform compute the true value from user-submitted data; the proposed privacy-preserving truth

---

[1] https://www.sohu.com/a/230282546_393779.

[2] https://www.sohu.com/a/250851584_119778?_f=index_pagerecom_7.

[3] https://www.chinanews.com.cn/gn/2020/08-31/9278816.shtml.

[4] https://www.sohu.com/a/412850764_99956743?_trans_=060005_xxhd.

[5] https://www.sohu.com/a/434247497_120893930.

discovery scheme with truth hiding helps the data requester obtain the true value from user-submitted data; the proposed privacy-preserving truth discovery scheme with task hiding helps the data requester obtain the true value from user-submitted data without compromising the task privacy. At the theoretical level, the proposed privacy-preserving content-based task allocation scheme implements task keyword matching which is resistant to chosen-plaintext attacks; the proposed privacy-preserving location-based task allocation scheme implements arbitrary geometric range queries which are resistant to collusive attacks by servers and users; the proposed privacy-preserving truth discovery scheme with truth transparency implements privacy-preserving user weight updates and task truth updates in a truth transparency scenario; the proposed privacy-preserving truth discovery scheme with truth hiding implements privacy-preserving user weight updates and task truth updates in a truth hiding scenario; the proposed privacy-preserving truth discovery scheme with task hiding implements privacy-preserving user weight updates and task truth updates in a task hiding scenario.

In summary, the research work in this book improves the security and effectiveness of task allocation and truth discovery in mobile crowdsensing applications, provides new ideas and methods for user privacy protection in MCS, and promotes the application of MCS in various fields such as environmental sensing, traffic supervision, health data collection, and difficult question answering, with certain academic and application values.

## 1.2   Mobile Crowdsensing

This section introduces the system model and security model of MCS. Specifically, this section first describes the system model of MCS by introducing the roles and functions of participating entities. Then the security model is described by analyzing the privacy threats in data collection and data processing and explaining the different attacks of each entity in MCS.

### 1.2.1   System Model of MCS

MCS takes advantage of users with mobile smart devices as the basic sensing unit and uses a mobile crowdsensing platform (e.g., cloud server) as an intermediary to connect the sensing user with the data requester, helping the data requester to complete large-scale, complex data sensing tasks [2, 3]. As shown in Fig. 1.1, the mobile crowdsensing system mainly includes the data requester, the mobile crowdsensing platform, and the sensing user. The roles and functions of each entity are defined as follows:

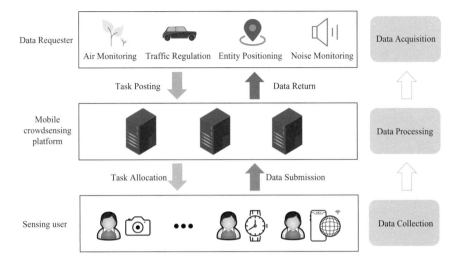

**Fig. 1.1** MCS system model

- Data requester: The data requester can be an entity such as a company, an organization, or an individual. The data requester has several data sensing tasks such as noise data collection, traffic information collection, air data collection, etc. To improve the efficiency of data collection and reduce the cost of data collection, the data requester issues data sensing tasks to the mobile crowdsensing platform and pays the sensing user and the mobile crowdsensing platform for the data they receive.
- Mobile crowdsensing platform: The mobile crowdsensing platform is usually a third-party platform operated by some companies or organizations (e.g., Baidu Cloud, Ali Cloud, Microsoft Cloud). The mobile crowdsensing platform recruits suitable sensing users for the data requester and is responsible for integrating and processing the data. Once the data has been processed, the mobile crowdsensing platform returns the final results to the data requester.
- Sensing user: The sensing user is usually an individual with a mobile smart device. When receiving a task from the mobile crowdsensing platform, the sensing user uses the GPS, camera, accelerometer, temperature sensor, air sensor, etc., in the mobile smart device to collect the corresponding data and upload it to the mobile crowdsensing platform.

In addition, to accommodate different applications or functions, existing MCS schemes also consider trusted centers [20, 21], multiple servers [22, 23], fog nodes [24–26], edge nodes [27, 28], etc., based on the above system framework.

## 1.2.2  Security Model of MCS

In MCS, the user is the subject of data collection. Thus, the data collected by the user is usually closely related to his/her personal information, such as locations, hobbies, beliefs, and trajectories. Other entities may infringe on the user's location, identity, personality, and data privacy after obtaining the user's data. For example, after acquiring the user's hobbies, the mobile crowdsensing platform may target advertisements to users to gain more financial benefits. More seriously, some malicious attackers may stalk, rob, or spy on users after obtaining private information related to their locations (e.g., home address, real-time location, life track), posing a threat to their property and lives. Therefore, compared with sensor-based wireless sensor networks, MCS faces serious security and privacy issues. Figure 1.2 gives an overview of the privacy threats faced by MCS during the data collection and processing phases.

Specifically, during the data collection phase, the mobile crowdsensing platform needs to allocate tasks of interest to the sensing user. However, suppose the user's task information is not protected. In that case, the mobile crowdsensing platform or other entity may infer the user's interests and further infer sensitive information such as identity, job, hobbies, political bias, and religious beliefs based on the tasks that interest the user. When task information is associated with location, mobile crowdsensing platforms or other entities may access the user's location information and further speculate on his/her home location, behavioral habits, life path, nature of work, and economic level. During the data processing phase, the data submitted by the user may reveal sensitive information such as the user's identity and location. User weights calculated by algorithms such as truth discovery and machine learning

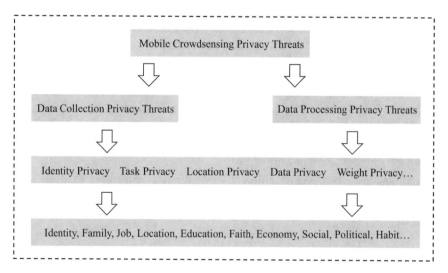

**Fig. 1.2**  Privacy threats in MCS

may reveal the user's information, to be used by mobile crowdsensing platforms or other entities to infer sensitive information such as the user's identity and occupation.

To retrieve user privacy, entities in MCS may launch various attacks based on their capabilities. This book adopts the security assumption that entities are "semi-trustworthy," i.e., all entities involved in a mobile crowdsensing application are assumed to honestly execute the designed protocols but may also try to retrieve other users' privacy information by various ways. To obtain the user's privacy, the following attacks are considered:

(1) Passive attack: A passive attack is an attack in which an adversary attempts to obtain the user's private information based on the data obtained in the protocol. For example, an external or internal non-mobile crowdsensing platform adversary may intercept communication data to infer the user's privacy. In contrast, a mobile crowdsensing platform may guess the privacy information hidden in the data submitted by the user. Passive attacks are cryptographically similar to ciphertext-only attacks.

(2) Active attack: In addition to the data directly available in the protocol, an adversary may attempt to obtain the users' private information based on extra information. An active attack is an attack that is initiated by an adversary based on the additional information obtained and it consists of the following main types:

- Known-context attack. An adversary may launch an attack based on the data distribution or statistical information related to the mobile crowdsensing application (e.g., the mobile crowdsensing platform may work with other platforms to obtain the frequency of occurrence of some data in the mobile crowdsensing application).
- Known-plaintext attack. An adversary may launch an attack based on some known plaintext-ciphertext pairs (e.g., a mobile crowdsensing platform may have some stale application-related plaintext-ciphertext datasets; an external or internal non-mobile crowdsensing platform adversary may learn the plaintext-ciphertext correspondence of some data by observing the location or preferences of some users and by listening to their communication channels).
- Chosen-plaintext attack. An adversary may launch an attack based on the chosen special plaintext message and obtain the corresponding ciphertext (e.g., an external or an internal non-mobile crowdsensing platform adversary may have some access to a smart card or application that stores the user's key. The adversary may be able to select a special plaintext message as input and observe the corresponding output; the mobile crowdsensing platform may conspire with some users to obtain a special plaintext-ciphertext pair).
- Collusion attack. An adversary may launch an attack based on some corrupted users (e.g., the mobile crowdsensing platform may conspire with some users to obtain information such as users' keys, uploaded data, and special plaintext-ciphertext pairs).

This book will set up the appropriate security models according to the needs of different applications and design privacy-preserving schemes under those security models. It should be noted that there are other attacks in MCS, such as witch attack [29, 30], denial of service attack [31, 32], and false data injection attack [33, 34]. These attacks are also important, but not discussed in this book.

## 1.3  The State of the Art and Trend of Privacy-Preserving

To achieve user privacy preservation during data collection and data processing in MCS, a great deal of research work has been carried out, and a series of effective privacy-preserving schemes have been designed. This section provides a brief overview of the works related to task allocation in the data collection and truth discovery in the data processing.

### 1.3.1  Privacy-Preserving Task Allocation

Task allocation based on task content and task allocation based on location are typical forms of allocation for mobile crowdsensing task allocation. This section provides a brief overview of these two forms of task allocation.

#### 1.3.1.1  Privacy-Preserving Content-Based Task Allocation

The most basic form of task allocation is task allocation based on task content in MCS, of which the basic principle is to find users who match the needs of the data requester based on their task content, improving the efficiency and accuracy of data collection. Many scholars have conducted research on task allocation based on task content, and a series of research results have been achieved. Among them, Yuen et al. [35] proposed a task recommendation framework based on Amazon's Mechanical Turk (MTurk) mobile crowdsensing platform, which models users' task preferences and then recommends tasks that they prefer to engage in based on the model. Feild et al. [36] proposed an allocation model based on task context semantics, aiming to improve the accuracy of task allocation. Aldahari et al. [37] studied a task allocation system based on user interests and skills for a multi-objective system. Their system considers matching the employer's needs with the user's interests and considers the economic and time costs required for task completion. In addition, their system can efficiently complete task allocation by designing a user interest matching algorithm.

As security and privacy issues have received increasing attention, many scholars have also proposed privacy-preserving task allocation schemes based on task content [20, 21, 38–43]. Specifically, Shu et al. [40] proposed a privacy-preserving task allocation scheme based on task content for mobile crowdsensing systems. Their scheme uses the ElGamal algorithm to encrypt the task information and match the user's interest with the data requester's task requirements in ciphertext form. Shu et al. [20] further proposed a privacy-preserving task allocation scheme based on [40]. Specifically, their scheme uses matrix decomposition to generate keys for the user, adopts polynomial functions and Veda's theorem to represent the user's task information, encrypts the task information using techniques such as secure kNN [44] and proxy re-encryption, and uses the nature of secure kNN techniques to compute the user's task matching score in ciphertext form. Due to the utilization of secure kNN techniques, their scheme can achieve high task matching performance; however, it is only resistant to known-plaintext attack. In addition, their scheme assumes that all users are fully trusted, which is a strong security assumption in practical applications. Thus, scheme [20] is difficult to resist in real-world application environments. Wu et al. [41] used blockchain technology [45, 46] to design a privacy-preserving task allocation algorithm for untrusted cloud servers. Tang et al. [42] and Xu et al. [43] designed task encryption and task matching protocols using bilinear mapping. Compared with scheme [20], schemes [42, 43] can achieve security under chosen-plaintext attack; however, due to the use of a time-consuming bilinear mapping algorithm, these two schemes are only suitable for smaller mobile crowdsensing scenarios. Ni et al. [21] used a random matrix multiplication technique to protect task information and implement task matching as a random matrix. Specifically, their scheme utilizes each element of the matrix to represent the task, employs a randomly generated matrix to protect the task matrix, and uses the nature of the matrix to achieve task matching between the data requester and the sensing user. Compared with the schemes [20, 42, 43], their scheme achieves a better balance between security and efficiency. However, the dimensionality of the matrix in the [21] is linearly related to the number of task types. Since there are usually a large number of task types in crowdsensing scenarios, [21] will impose a large computational and communication overhead on the user and the mobile crowdsensing platform in practice.

### 1.3.1.2    Privacy-Preserving Location-Based Task Allocation

Most of the tasks in MCS are closely related to the user's location. Therefore, task allocation based on location [47] has also attracted a great deal of scholarly attention. Therefore, how to achieve location privacy protection for users is an important research component of current location-based task allocation. In recent years, many scholars have conducted an in-depth research on privacy-preserving task allocation based on geometric range, and a series of original research results have been achieved. For example, Liu et al. [48] used the Paillier homomorphic encryption algorithm [49] to encrypt the user's latitude and longitude coordinates and find the user closest to the task required by the data requester in the ciphertext form. Huang et al. [50], Wang et al. [51], and Zhang et al. [52] used differential privacy techniques [53] to protect the user's location information and allocated the closest task to the user while ensuring the security and availability of the user's location. However, the above schemes can only support task allocation within a circular range or at the closest distance. In practical mobile crowdsensing scenarios, the geometric ranges requested by data requesters are often irregular due to factors such as building distribution, geography, and task requirements, so there is a need to investigate privacy-preserving task allocation schemes based on arbitrary geometric range queries in MCS.

A number of current scholars have proposed schemes to support arbitrary geometric range queries [54–57]. Xu et al. [54] proposed a location query scheme based on secure kNN techniques to support arbitrary geometric ranges. Their scheme uses a polynomial fitting function to fit an arbitrary geometric range and uses secure kNN techniques to achieve geometric location and query range protection. Further, their scheme achieves nonlinear retrieval efficiency using order-preserving encryption. Wang et al. [57] encoded the user's location using quadtrees and Gray codes and used the SHVE encryption algorithm [58] to encrypt the geometric locations and query ranges. However, it should be noted that most of the above schemes are based on symmetric encryption and are unsuitable for mobile crowdsensing applications. Based on public keys, Alamer et al. [59] designed a privacy-preserving user retrieval scheme that can support arbitrary geometric range queries using bilinear mapping. The scheme of Ni et al. [21] based on the random matrix multiplication technique can also be extended to support arbitrary geometric range queries. However, the above two schemes encrypt the geometric range in chunks, and the computational overhead of data encryption and data retrieval is linearly related to the geometric chunk size. When the query range is large, both schemes will impose a large computational overhead on the user and the mobile crowdsensing platform. In addition, most existing schemes are not resistant to collusive attacks by the mobile crowdsensing platform and the user. In practice, the mobile crowdsensing platform may obtain some plaintext and ciphertext information about the geometric range through collusion. When the user updates the location, the mobile crowdsensing platform uses the existing geometric range information to determine whether the user's updated location is within the existing geometric range.

## *1.3.2  Privacy-Preserving Truth Discovery*

Due to the varying quality and distribution of sensing devices, the different identities, background knowledge, and behavioral habits of device holders, the quality of data collected by sensing users often differs significantly. To calculate the true sensing data from user-submitted data, there have been many data reliability estimation methods, such as data mining [60], machine learning [61], fuzzy logic [62], and truth discovery [63]. Compared with other data reliability estimation methods, truth discovery is widely used in MCS as it can efficiently compute the true outcome of each task from multi-task data submitted by users without any prior knowledge or background knowledge [64–67]. The basic principles of truth discovery are (1) if a user's data is close to the true outcome (i.e., the truth), then that user is given a higher weight, and (2) if a user has a higher weight, then that user's data is given a higher weight in the calculation of the truth. By performing the above two steps iteratively, the truth of the task is obtained when the iteration terminates.

Several scholars have proposed a series of privacy-preserving truth discovery schemes to protect the privacy of users' data and weights in recent years. Miao et al. [68] proposed the first privacy-preserving truth discovery scheme. Specifically, their scheme uses a homomorphic encryption algorithm to encrypt the user's data and iteratively update the true value of the task through the interaction between the cloud server and the user. To prevent the cloud server or a single user from restoring the perceptual data and weights, the scheme splits the decrypted private key, distributes them to $p$ users, and randomly selects $t$ users to restore the aggregated results. Furthermore, their scheme uses the MapReduce framework to design a parallel privacy-preserving truth discovery scheme to improve the system's operational efficiency. Although this scheme is good at protecting the privacy of users' data and weights, it has a large computational and communication overhead due to the use of time-consuming cryptographic algorithms. To achieve efficient privacy-preserving truth discovery, Xu et al. [69] designed an efficient privacy-preserving truth discovery scheme using a random perturbation-based homomorphic cryptographic aggregation protocol [70]. Specifically, their scheme allocates a random data to each user participating in the sensing task and sends the sum of all users' random data to the mobile crowdsensing platform. The user adds random data to the sensing data and sends it to the mobile crowdsensing platform, where the results of the perturbations from all users are aggregated. The sum of the random data is subtracted to obtain the true aggregated value. This scheme has high computational and communication efficiency due to the utilization of random data perturbation operations. However, because all users need to submit data on time to remove the random data perturbation, the scheme has poor system fault tolerance, i.e., once a user fails to upload data on time, the scheme will get an incorrect truth. Therefore, to achieve efficient and secure truth discovery and improve the system's fault tolerance, Xu et al. [71] designed a single-server-based privacy-preserving truth discovery scheme using the Shamir key-sharing protocol [72]. Similarly, Zheng et al. [73] also designed a single-server and two-server privacy-preserving truth discovery scheme using the key-sharing protocol. The experimental results show that these schemes have significant computational and communication improvements over existing schemes [68]. However, considering that users in MCS may engage or exit the system frequently due to network conditions or mobility, the secret sharing-based scheme relies on frequent interactions between servers and users to update the keys, increasing the system's redundancy.

To reduce the interaction between servers and users, Miao et al. [23] proposed a privacy-preserving truth discovery scheme based on two non-collusive servers. Specifically, each user generates random data to perturb the sensing data and sends the perturbed results and random data to each of the two cloud servers. The two cloud servers then interactively update the true values and send them to each user, who updates their weights based on the true values. However, since users are involved in calculating their weights, which are often closely related to their incentives and rewards, they may tamper with their weights' values to obtain more rewards. To further reduce the computational and communication overhead of user participation in the system and to prevent users from tampering with their weights, privacy-preserving truth discovery schemes based on two non-colluding servers where users do not need to participate in the truth discovery iterative process have also been proposed [23, 74, 75]. Among them, Miao et al. [23] proposed the first privacy-preserving truth discovery scheme in which users do not need to participate in the truth discovery iterative process. Specifically, their scheme uses two non-colluding cloud servers as the mobile crowdsensing platform and uses data perturbation techniques to add noise to the user's data for protecting the privacy of the sensing data. To perform truth discovery without compromising data privacy, their scheme uses the Paillier homomorphic encryption algorithm to update the true values and weights through the interactions between the two cloud servers. However, their scheme fails to protect the privacy of the user's weight.

Tang et al. [74] constructed a privacy-preserving truth discovery scheme on a two-server platform using garbled circuits [76] and inadvertent transmission [77] without user participation in the truth discovery iterative process. In this scheme, the user first generates a series of perturbation data and then sends the results of the perturbation along with the perturbation data to two separate servers. After that, one server generates a garbled circuit associated with the truth discovery and sends the garbled circuit along with the garbled perturbation data to the other server. Similarly, based on two servers, Cai et al. [78] used Beaver's multiplicative triples [79] and oblivious transfer (OT) to implement a privacy-preserving truth discovery framework that does not require users to participate in the truth discovery iteration process. In addition, their scheme uses blockchain and smart contracts to implement a payment mechanism that provides rewards based on the quality of the user's data. However, both schemes take much longer to generate the garbled circuits and incur greater computational and communication overheads. By using the Paillier homomorphic encryption algorithm, Tang et al. [75] implemented a privacy-preserving truth discovery algorithm based on two servers to protect user data and weights. However, their scheme requires more interactions between the two cloud servers to achieve privacy-preserving truth and weight updates, which imposes a large computational and communication overhead on the system. Tang et al. [80] implement a $k$-anonymous truth discovery scheme that protects the user's identity privacy and prevents attackers from determining the origin of the data; however, the scheme does not protect the user's data and weight privacy. In practice, the mobile crowdsensing platform may use the user's data for its further financial gain. In addition, most of the above schemes do not consider the scenario where the data requester is involved in truth discovery, and truth privacy also needs to be protected. In the real-world mobile crowdsensing scenario, the data requester needs to pay for the truth, and the transparency of the truth to the user or the mobile crowdsensing platform would compromise the privacy and financial interests of the data requester.

   In summary, although many scholars are now focusing on privacy-preserving task allocation in MCS data collection and privacy-preserving truth discovery in data processing, many original research results have been achieved. The current research still suffers from insufficient privacy-preserving capabilities, high computational and communication overheads, and relatively unified support functions. With the widespread use of MCS in social production and life and the increased awareness of security and privacy protection, it is necessary to follow the development of MCS and investigate a more secure and efficient user privacy protection scheme applicable to the mobile crowdsensing environment.

## 1.4   Organization of the Book

This book is dedicated to providing comprehensive privacy protection for users in MCS data collection and processing phases. In the data collection phase, it proposes a privacy-preserving content-based task allocation scheme and a privacy-preserving location-based task allocation scheme; in the data collection phase, it proposes a privacy-preserving truth discovery scheme with truth transparency, a privacy-preserving truth discovery scheme with truth hiding, and a privacy-preserving truth discovery scheme with task hiding. The research content and main contributions of this book are summarized as follows:

(1) Privacy-preserving content-based task allocation: For the content-based task allocation scenario, this book investigates efficient and accurate task allocation while protecting the privacy of sensing users' tasks and data requester's query. The main contributions of this research include:

- Based on random matrix multiplication techniques and polynomial functions, a privacy-preserving content-based task allocation scheme (PPTA) is proposed. Specifically, PPTA uses polynomial functions to represent task and query information, uses random matrix multiplication techniques to protect the privacy of task and query, and uses polynomial functions and matrix properties to find sensing users in a random matrix that matches the task requirements of the data requester.
- PPTA is extended to support further privacy-preserving task allocation, threshold task allocation, task allocation with access control support, and task reduction in mobile crowdsensing task allocation.
- The security and privacy analysis proves that PPTA can effectively protect the privacy of users' tasks and queries. Experiments based on a mobile crowdsensing environment validate the efficiency of PPTA in the data encryption and task retrieval phases.

(2) Privacy-preserving location-based task allocation: For the location-based task allocation scenario, this chapter investigates efficient and accurate task alloca-

tion while protecting the privacy of the sensing user's location and the privacy of the data requester's query. The main contributions of this research include:

- Based on the two-server model, the privacy-preserving task allocation scheme GPTA-L is proposed to support arbitrary geometric range queries, which uses matrix decomposition to generate keys, polynomial fitting techniques to generate query ranges, and random matrix multiplication techniques to protect location and query privacy. Using polynomial fitting functions and matrix properties, GPTA-L enables the mobile crowdsensing platform to find sensing users within the data requester's geometric query range without knowing their specific location and query information.
- To further improve retrieval efficiency, GPTA-F, a scheme that enables nonlinear retrieval efficiency, is designed based on GPTA-L to study the historical query behavior of data requesters. Specifically, GPTA-F sets a label for a data requester's historical query and determines whether the search results under that label are within the new query by matching the new query to the label.
- Security and privacy analysis demonstrates that GPTA-L and GPTA-F can effectively protect the privacy of users' location and queries. Experiments based on the mobile crowdsensing environment have verified that GPTA-L and GPTA-F achieve high computational and communication efficiency.

(3) Privacy-preserving truth discovery with truth transparency: For the truth discovery scenario with truth transparency, this chapter investigates the realization of efficient truth discovery under the premise of protecting sensing users' data and weight privacy. The main contributions of this research include:

- For the scenario where the user's location is relatively fixed and the network condition is good, the scheme RPTD-I is proposed, in which the sensing user is involved in the iterative process of truth discovery, and techniques such as homomorphic encryption algorithms and superlinear sequences are used to ensure the privacy and efficiency requirements of the scheme.
- For the scenario where the user moves frequently, the proposed scheme RPTD-II uses homomorphic encryption algorithms and data perturbation techniques to protect the privacy of the user's data and weights. It transfers the user's computational operations during the truth discovery iteration to the cloud server, significantly reducing the user's computational and communication overhead.
- The security and privacy analysis demonstrates that RPTD-I and RPTD-II effectively protect the user's data and weight privacy. Experiments based on the mobile crowdsensing environment validate that RPTD-I and RPTD-II have high computational and communication efficiency.

(4) Privacy-preserving truth discovery with truth hiding: For the truth discovery scenario with truth hiding, this chapter investigates efficient truth discovery while protecting the privacy of sensing user's data and weight and the data requester's truth. The main contributions of this research include:

- The privacy-preserving truth discovery scheme SATE, which does not require the user to participate in the truth discovery iteration process, is proposed to provide full privacy protection for the truth of the data requester and the data and weights of the sensing user.
- Designing lightweight data perturbation algorithms to protect the data privacy of sensing users. SATE achieves lower computational and communication overheads compared with existing solutions on the sensing user side. In addition, based on data perturbation techniques and public-key cryptosystems that support multi-party collaborative decryption, the design of privacy-preserving truth discovery algorithms allows the cloud server to implement privacy-preserving truth updates and weight updates without user participation.
- Experiments based on the mobile crowdsensing environment validate that SATE has a low computational and communication overhead on both the user and mobile crowdsensing platforms.

(5) Privacy-preserving truth discovery with task hiding: For the truth discovery scenario with task hiding, this chapter investigates efficient truth discovery while protecting the privacy of the sensing user' data and the data requester's tasks. The main contributions of this research include:

- Studying the challenging problems of truth discovery in crowdsensing applications and identifying its utility and security requirements by analyzing users' privacy needs.
- The lightweight privacy-preserving truth discovery scheme, LPTD, is proposed for crowdsensing systems, which uses data perturbation techniques and the properties of matrix multiplication to enable correct computation of the final truth from conflicting responses while protecting data and task privacy.
- The security and privacy analysis demonstrates that LPTD effectively protects the user's data and the data requester's tasks. Experiments based on the mobile crowdsensing environment validate that LPTD has high accuracy and efficiency.

This book proposes several user privacy protection schemes for data collection and processing. The proposed schemes effectively protect the privacy of users' tasks, locations, data, weights, and truth with low computational and communication overheads and have some academic and practical application value.

The book is organized as follows: Chap. 1 first introduces the research background and significance of MCS, then introduces the system and security model of MCS, reviews the research work related to task allocation and truth discovery in the data collection and data processing phases, summarizes the research content and contributions of this chapter based on the above, and finally gives the organization of this chapter. Chapter 2 introduces the basic knowledge required for this book. Chapter 3 investigates the privacy-preserving content-based task allocation in the data collection phase and proposes a privacy-preserving content-based task allocation scheme, PPTA, to address the problems of single-function support and poor task matching efficiency of existing privacy-preserving schemes. Chapter 4 investigates location-based task allocation in the data collection phase and proposes GPTA, a privacy-preserving location-based task allocation scheme that supports arbitrary geometric range queries in response to restricted query range and poor location retrieval efficiency of existing privacy-preserving solutions. Chapter 5 investigates the privacy-preserving truth discovery in the data processing phase in the truth transparency scenario and proposes an efficient privacy-preserving truth discovery scheme with truth transparency, RPTD, because of the weak privacy protection capability and poor efficiency of existing privacy-preserving schemes. Chapter 6 investigates privacy-preserving truth discovery in the data processing phase in the truth hiding scenario and proposes an efficient privacy-preserving truth discovery scheme SATE with truth hiding because of the difficulties of existing privacy-preserving schemes to effectively protect truth privacy and the high overheads on the sensing user side. Chapter 7 investigates privacy-preserving truth discovery in the data processing phase in the task hiding scenario and proposes an efficient privacy-preserving truth discovery scheme with task hiding, LPTD, to effectively protect data privacy and task privacy and achieve high efficiency on the sensing user side, in response to the difficulties of existing privacy-preserving schemes. Finally, the book summarizes the research work and discusses future research directions. The organization of this chapter and the logical connection of each chapter are shown in Fig. 1.3.

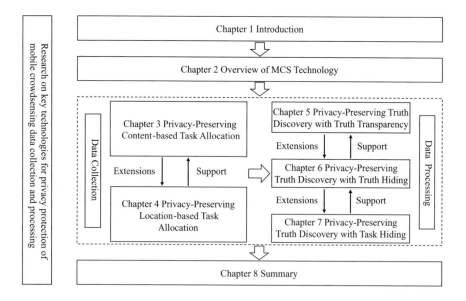

**Fig. 1.3** Chapter organization of this book

# References

1. Ganti, R.K., Ye, F., Lei, H.: Mobile crowdsensing: current state and future challenges. IEEE Commun. Mag. **49**(11), 32–39 (2011)
2. Guo, B., Yu, Z., Zhou, X., Zhang, D.: From participatory sensing to mobile crowd sensing. In: 2014 IEEE International Conference on Pervasive Computing and Communication Workshops (PERCOM WORKSHOPS), pp. 593–598. IEEE (2014)
3. Guo, B., Wang, Z., Yu, Z., Wang, Y., Yen, N.Y., Huang, R., Zhou, X.: Mobile crowd sensing and computing: The review of an emerging human-powered sensing paradigm. ACM Comput. Surv. **48**(1), 1–31 (2015)
4. Liu, J., Shen, H., Narman, H.S., Chung, W., Lin, Z.: A survey of mobile crowdsensing techniques: a critical component for the internet of things. ACM Trans. Cyber-Phys. Syst. **2**(3), 1–26 (2018)
5. Capponi, A., Fiandrino, C., Kantarci, B., Foschini, L., Kliazovich, D., Bouvry, P.: A survey on mobile crowdsensing systems: Challenges, solutions, and opportunities. IEEE Commun. Surv. Tutor. **21**(3), 2419–2465 (2019)
6. Liu, Y., Yu, Z., Guo, B., Han, Q., Su, J., Liao, J.: Crowdos: a ubiquitous operating system for crowdsourcing and mobile crowd sensing. IEEE Trans. Mob. Comput. **21**(3): 878–894 (2020)
7. Tang, L., Yang, X., Dong, Z., Li, Q.: CLRIC: Collecting lane-based road information via crowdsourcing. IEEE Trans. Intell. Transport. Syst. **17**(9), 2552–2562 (2016)
8. Teng, X., Guo, D., Guo, Y., Zhou, X., Ding, Z., Liu, Z.: IONavi: An indoor-outdoor navigation service via mobile crowdsensing. ACM Trans. Sensor Netw. **13**(2), 1–28 (2017)
9. Ni, J., Zhang, K., Yu, Y., Lin, X., Shen, X.: Privacy-preserving smart parking navigation supporting efficient driving guidance retrieval. IEEE Trans. Vehic. Technol. **67**(7), 6504–6517 (2018)
10. Zappatore, M., Longo, A., Bochicchio, M.A.: Using mobile crowd sensing for noise monitoring in smart cities. In: 2016 International Multidisciplinary Conference on Computer and Energy Science (Splitech), pp. 1–6. IEEE (2016)
11. Pan, Z., Yu, H., Miao, C., Leung, C.: Crowdsensing air quality with camera-enabled mobile devices. In: Proceedings of the AAAI Conference on Artificial Intelligence, vol. 31, pp. 4728–4733 (2017)
12. Ma, R., Chen, X., Liu, H., Jinbo, X.: Research on user privacy measurement and privacy protection in mobile crowd sensing (in Chinese). Netinf. Secur. **18**(8), 64 (2018)
13. Xiong, J., Ma, R., Niu, B., Guo, Y., Lin, L.: Privacy protection incentive mechanism based on user alliance matching in mobile crowd sensing (in Chinese). J. Comput. Res. Develop. **55**(7), 1359–1370 (2018)
14. Xiong, J., Ma, R., Chen, L., Tian, Y., Li, Q., Liu, X., Yao, Z.: A personalized privacy protection framework for mobile crowdsensing in IIoT. IEEE Trans. Ind. Inf. **16**(6), 4231–4241 (2019)
15. Wang, Z., Pang, X., Hu, J., Liu, W., Wang, Q., Li, Y., Chen, H.: When mobile crowdsensing meets privacy. IEEE Commun. Mag. **57**(9), 72–78 (2019)
16. Zhao, B., Tang, S., Liu, X., Zhang, X.: PACE: privacy-preserving and quality-aware incentive mechanism for mobile crowdsensing. IEEE Trans. Mob. Comput. **20**(5), 1924–1939 (2020)
17. Wang, Z., Li, J., Hu, J., Ren, J., Li, Z., Li, Y.: Towards privacy-preserving incentive for mobile crowdsensing under an untrusted platform. In: IEEE INFOCOM 2019-IEEE Conference on Computer Communications, pp. 2053–2061. IEEE (2019)
18. Fang, W., Zhou, C., Sun, S.: Research on task assignment in mobile crowd sensing (in Chinese). Appl. Res. Comp. **35**(11), 3206–3212 (2018)
19. Ma, R., Meng, X.: Research on truth discovery method based on data source classification reliability (in Chinese). J. Comput. Res. Develop. **52**(9), 1931–1940 (2015)
20. Shu, J., Jia, X., Yang, K., Wang, H.: Privacy-preserving task recommendation services for crowdsourcing. IEEE Trans. Serv. Comput. **14**(1), 235–247 (2018)
21. Ni, J., Zhang, K., Xia, Q., Lin, X., Shen, X.S.: Enabling strong privacy preservation and accurate task allocation for mobile crowdsensing. IEEE Trans. Mob. Comput. **19**(6), 1317–

1331 (2019)

22. Liu, B., Chen, L., Zhu, X., Zhang, Y., Zhang, C., Qiu, W.: Protecting location privacy in spatial crowdsourcing using encrypted data. In: Advances in Database Technology-EDBT (2017)

23. Miao, C., Su, L., Jiang, W., Li, Y., Tian, M.: A lightweight privacy-preserving truth discovery framework for mobile crowd sensing systems. In: IEEE INFOCOM 2017-IEEE Conference on Computer Communications, pp. 1–9. IEEE (2017)

24. Ni, J., Zhang, A., Lin, X., Shen, X.S.: Security, privacy, and fairness in fog-based vehicular crowdsensing. IEEE Commun. Mag. **55**(6), 146–152 (2017)

25. Belli, D., Chessa, S., Kantarci, B., Foschini, L.: Toward fog-based mobile crowdsensing systems: state of the art and opportunities. IEEE Commun. Mag. **57**(12), 78–83 (2019)

26. Basudan, S., Lin, X., Sankaranarayanan, K.: A privacy-preserving vehicular crowdsensing-based road surface condition monitoring system using fog computing. IEEE Int. Things J. **4**(3), 772–782 (2017)

27. Marjanović, M., Antonić, A., Žarko, I.P.: Edge computing architecture for mobile crowdsensing. IEEE Access **6**, 10662–10674 (2018)

28. Ma, L., Liu, X., Pei, Q., Xiang, Y.: Privacy-preserving reputation management for edge computing enhanced mobile crowdsensing. IEEE Trans. Serv. Comput. **12**(5), 786–799 (2018)

29. Chang, S.-H., Chen, Z.-R.: Protecting mobile crowd sensing against sybil attacks using cloud based trust management system. Mobile Inf. Syst.**2016**, 1–10 (2016)

30. Yun, J., Kim, M.: Sybileye: observer-assisted privacy-preserving sybil attack detection on mobile crowdsensing. Information **11**(4), 198 (2020)

31. Jia, B., Liang, Y.: Anti-d chain: a lightweight ddos attack detection scheme based on heterogeneous ensemble learning in blockchain. China Commun. **17**(9), 11–24 (2020)

32. Huang, J., Kong, L., Dai, H.-N., Ding, W., Cheng, L., Chen, G., Jin, X., Zeng, P.: Blockchain-based mobile crowd sensing in industrial systems. IEEE Trans. Ind. Inf. **16**(10), 6553–6563 (2020)

33. Zhou, T., Cai, Z., Xu, M., Chen, Y.: Leveraging crowd to improve data credibility for mobile crowdsensing. In: IEEE Symposium on Computers and Communication (ISCC), pp. 561–568. IEEE (2016)

34. Li, X., Xie, K., Wang, X., Xie, G., Xie, D., Li, Z., Wen, J., Diao, Z., Wang, T.: Quick and accurate false data detection in mobile crowd sensing. IEEE/ACM Trans. Netw. **28**(3), 1339–1352 (2020)

35. Yuen, M.-C., King, I., Leung, K.-S.: Taskrec: a task recommendation framework in crowd-sourcing systems. Neural Process. Lett. **41**(2), 223–238 (2015)

36. Feild, H., Allan, J.: Task-aware query recommendation. In: Proceedings of the 36th International ACM SIGIR Conference on Research and Development in Information Retrieval, pp. 83–92 (2013)

37. Aldahari, E., Shandilya, V., Shiva, S.: Crowdsourcing multi-objective recommendation system. In: Companion Proceedings of the The Web Conference 2018, pp. 1371–1379 (2018)

38. Gong, Y., Guo, Y., Fang, Y.: A privacy-preserving task recommendation framework for mobile crowdsourcing. In: 2014 IEEE Global Communications Conference, pp. 588–593. IEEE (2014)

39. Ni, J., Zhang, K., Yu, Y., Lin, X., Shen, X.S.: Providing task allocation and secure deduplication for mobile crowdsensing via fog computing. IEEE Trans. Dependable Secure Comput. **17**(3), 581–594 (2020)

40. Shu, J., Jia, X.: Secure task recommendation in crowdsourcing. In: 2016 IEEE Global Communications Conference (GLOBECOM), pp. 1–6. IEEE (2016)

41. Wu, Y., Tang, S., Zhao, B., Peng, Z.: BPTM: Blockchain-based privacy-preserving task matching in crowdsourcing. IEEE Access **7**, 45605–45617 (2019)

42. Tang, W., Zhang, K., Ren, J., Zhang, Y., Shen, X.S.: Privacy-preserving task recommendation with win-win incentives for mobile crowdsourcing. Inf. Sci. **527**, 477–492 (2020)

43. Xu, J., Cui, B., Shi, R., Feng, Q.: Outsourced privacy-aware task allocation with flexible expressions in crowdsourcing. Future Gener. Comput. Syst. **112**, 383–393 (2020)

44. Wong, W.K., Cheung, D.W.-L., Kao, B., Mamoulis, N.:   Secure kNN computation on encrypted databases. In: Proceedings of the 2009 ACM SIGMOD International Conference on Management of Data, pp. 139–152 (2009)
45. Yuan, Y., Wang, F.: Current situation and prospects of blockchain technology development (in Chinese). Acta Autom. Sinica **42**(4), 481–494 (2016)
46. Zhu, L., Gao, F., Shen, M., Li, Y., Zheng, B., W.-Z. Mao, Honggao: Overview of blockchain privacy protection research (in Chinese). J. Comput. Res. Develop. **54**(10), 2170–2186 (2017)
47. Zhu, H., Lu, R., Huang, C., Chen, L., Li, H.: An efficient privacy-preserving location-based services query scheme in outsourced cloud. IEEE Trans. Vehic. Technol. **65**(9), 7729–7739 (2015)
48. Liu, A., Li, Z.X., Liu, G.F., Zheng, K., Zhang, M., Li, Q., Zhang, X.: Privacy-preserving task assignment in spatial crowdsourcing. J. Comput. Sci. Technol. **32**, 905–918 (2017)
49. Paillier, P.:   Public-key cryptosystems based on composite degree residuosity classes. In: International Conference on the Theory and Applications of Cryptographic Techniques, pp. 223–238. Springer (1999)
50. Huang, C., Lu, R., Zhu, H., Shao, J., Alamer, A., Lin, X.: EPPD: efficient and privacy-preserving proximity testing with differential privacy techniques. In: 2016 IEEE International Conference on Communications (ICC), pp. 1–6. IEEE (2016)
51. Wang, L., Yang, D., Han, X., Wang, T., Zhang, D., Ma, X.: Location privacy-preserving task allocation for mobile crowdsensing with differential geo-obfuscation. In: Proceedings of the 26th International Conference on World Wide Web, pp. 627–636 (2017)
52. Zhang, X., Ding, J., Li, X., Yang, T., Wang, J., Pan, M.: Mobile crowdsensing task allocation optimization with differentially private location privacy. In: ICC 2020–2020 IEEE International Conference on Communications (ICC), pp. 1–6. IEEE (2020)
53. Dwork, C.: Differential privacy: A survey of results. In: International Conference on Theory and Applications of Models of Computation, pp. 1–19. Springer (2008)
54. Xu, G., Li, H., Dai, Y., Yang, K., Lin, X.: Enabling efficient and geometric range query with access control over encrypted spatial data. IEEE Trans. Inf. Forensics Secur. **14**(4), 870–885 (2018)
55. Wang, B., Li, M., Wang, H.: Geometric range search on encrypted spatial data. IEEE Trans. Inf. Forensics Secur. **11**(4), 704–719 (2015)
56. Wang, B., Li, M., Xiong, L.: Fastgeo: Efficient geometric range queries on encrypted spatial data. IEEE Trans. Dependable Secure Comput. **16**(2), 245–258 (2017)
57. Wang, X., Ma, J., Liu, X., Deng, R.H., Miao, Y., Zhu, D., Ma, Z.; Search me in the dark: Privacy-preserving boolean range query over encrypted spatial data. In: IEEE INFOCOM 2020-IEEE Conference on Computer Communications, pp. 2253–2262. IEEE (2020)
58. Lai, S., Patranabis, S., Sakzad, A., Liu, J.K., Mukhopadhyay, D., Steinfeld, R., Sun, S.F., Liu, D., Zuo, C.:   Result pattern hiding searchable encryption for conjunctive queries. In: Proceedings of the 2018 ACM SIGSAC Conference on Computer and Communications Security, pp. 745–762 (2018)
59. Alamer, A., Ni, J., Lin, X., Shen, X.: Location privacy-aware task recommendation for spatial crowdsourcing. In: 2017 9th International Conference on Wireless Communications and Signal Processing (WCSP), pp. 1–6. IEEE (2017)
60. Ouyang, Y., Yan, T., Wang, G.: Crowdmi: scalable and diagnosable mobile voice quality assessment through wireless analytics. IEEE Int. Things J. **2**(4), 287–294 (2015)
61. Liu, S., Zheng, Z., Wu, F., Tang, S., Chen, G.: Context-aware data quality estimation in mobile crowdsensing. In: IEEE INFOCOM 2017-IEEE Conference on Computer Communications, pp. 1–9. IEEE (2017)
62. An, J., Liang, D., Gui, X., Yang, H., Gui, R., He, X.: Crowdsensing quality control and grading evaluation based on a two-consensus blockchain. IEEE Int. Things J. **6**(3), 4711–4718 (2018)

63. Yang, S., Wu, F., Tang, S., Gao, X., Yang, B., Chen, G.: On designing data quality-aware truth estimation and surplus sharing method for mobile crowdsensing. IEEE J. Select. Areas Commun. **35**(4), 832–847 (2017)
64. Li, Q., Li, Y., Gao, J., Su, L., Zhao, B., Demirbas, M., Fan, W., Han, J.: A confidence-aware approach for truth discovery on long-tail data. Proc. VLDB Endowment **8**(4), 425–436 (2014)
65. Li, Q., Li, Y., Gao, J., Zhao, B., Fan, W., Han, J.: Resolving conflicts in heterogeneous data by truth discovery and source reliability estimation. In: Proceedings of the 2014 ACM SIGMOD International Conference on Management of Data, pp. 1187–1198 (2014)
66. Li, X., Dong, X. L., Lyons, K., Meng, W., Srivastava, D.: Truth finding on the deep web: Is the problem solved? Preprint arXiv:1503.00303 (2015)
67. Li, Y., Li, Q., Gao, J., Su, L., Zhao, B., Fan, W., Han, J.: On the discovery of evolving truth. In: Proceedings of the 21th ACM SIGKDD International Conference on Knowledge Discovery and Data Mining, pp. 675–684 (2015)
68. Miao, C., Jiang, W., Su, L., Li, Y., Guo, S., Qin, Z., Xiao, H., Gao, J., Ren, K.: Cloud-enabled privacy-preserving truth discovery in crowd sensing systems. In: Proceedings of the 13th ACM Conference on Embedded Networked Sensor Systems, pp. 183–196 (2015)
69. Xu, G., Li, H., Tan, C., Liu, D., Dai, Y., Yang, K.: Achieving efficient and privacy-preserving truth discovery in crowd sensing systems. Comput. Secur. **69**, 114–126 (2017)
70. Zhou, J., Cao, Z., Dong, X., Lin, X.: Security and privacy in cloud-assisted wireless wearable communications: Challenges, solutions, and future directions. IEEE Wirel. Commun. **22**(2), 136–144 (2015)
71. Xu, G., Li, H., Liu, S., Wen, M., Lu, R.: Efficient and privacy-preserving truth discovery in mobile crowd sensing systems. IEEE Trans. Vehic. Technol. **68**(4), 3854–3865 (2019)
72. Shamir, A.: How to share a secret. Commun. ACM **22**(11), 612–613 (1979)
73. Zheng, Y., Duan, H., Yuan, X., Wang, C.: Privacy-aware and efficient mobile crowdsensing with truth discovery. IEEE Trans. Dependable Secure Comput. **17**(1), 121–133 (2017)
74. Tang, X., Wang, C., Yuan, X., Wang, Q.: Non-interactive privacy-preserving truth discovery in crowd sensing applications. In: IEEE INFOCOM 2018-IEEE Conference on Computer Communications, pp. 1988–1996. IEEE (2018)
75. Tang, J., Fu, S., Xu, M., Luo, Y., Huang, K.: Achieve privacy-preserving truth discovery in crowdsensing systems. In: Proceedings of the 28th ACM International Conference on Information and Knowledge Management, pp. 1301–1310 (2019)
76. Huang, Y., Evans, D., Katz, J., Malka, L.: Faster secure { Two-Party } computation using garbled circuits. In: 20th USENIX Security Symposium (USENIX Security 11) (2011)
77. Rabin, M.O.: How to exchange secrets with oblivious transfer. Cryptology ePrint Archive (2005)
78. Cai, C., Zheng, Y., Wang, C.: Leveraging crowdsensed data streams to discover and sell knowledge: A secure and efficient realization. In: 2018 IEEE 38th International Conference on Distributed Computing Systems (ICDCS), pp. 589–599. IEEE (2018)
79. Beaver, D.: Efficient multiparty protocols using circuit randomization. In: Annual International Cryptology Conference, pp. 420–432. Springer (1991)
80. Tang J., Fu S., Liu X., Luo Y., Xu M.: Achieving privacy-preserving and lightweight truth discovery in mobile crowdsensing. IEEE Trans. Knowl. Data Eng. **34**(11), 5140–5153 (2021)

# Chapter 2
# Overview of MCS Technology

**Abstract** This book studies user privacy issues in data collection and data processing, which mainly involves the privacy-preserving content-based task allocation, the privacy-preserving location-based task allocation, the privacy-preserving truth discovery with truth transparency, and the privacy-preserving truth discovery with truth hiding. To realize the above schemes, the basic knowledge used in this chapter includes polynomial function, secure hash function, searchable encryption, asymmetric scalar-product-preserving encryption, polynomial fitting, truth discovery, and public-key cryptosystem supporting distributed decryption. The relevant basic knowledge is introduced as follows.

## 2.1 Preliminary of Privacy-Preserving Data Collection Techniques

### 2.1.1 Polynomial Function

Given a polynomial function $f(x) = a_n x^n + a_{n-1} x^{n-1} + \cdots + a_1 x + a_0$ with the highest power of $n$, where the coefficient of $x^i$ is $a_i$, $i \in [0, n]$, and $a_n \neq 0$, according to the basic theorem of algebra,

$$
\begin{aligned}
f(x) &= a_n x^n + a_{n-1} x^{n-1} + \cdots + a_1 x + a_0 \\
&= a_n (x - x_n)(x - x_{n-1}) \cdots (x - x_1),
\end{aligned}
\tag{2.1}
$$

where $(x_1, x_2, \ldots, x_n)$ is the root of polynomial function $f(x)$. According to Weida's theorem, the $n - k$th coefficient of $f(x)$, i.e., $a_{n-k} (k \in [0, n])$, can be calculated as

$$
\sum_{1 \leq i_1 < i_2 < \cdots < i_k \leq n} \left( \prod_{j=1}^{k} x_{i_j} \right) = (-1)^k \frac{a_{n-k}}{a_n}.
\tag{2.2}
$$

In the privacy-preserving content-based task allocation designed in this book, a polynomial function is used to judge whether a task $x_i$ belongs to the task set $\{x_1, x_2, \ldots, x_n\}$. That is, suppose $\{\texttt{Task}(m) \rightarrow x_m\}_{m=1}^n$ is the required task set, and $f(x) = a_n(x - x_n)(x - x_{n-1}) \cdots (x - x_1)$ is the corresponding polynomial function. If $f(x_t) = 0$, the task $\texttt{Task}(t) \rightarrow x_t$ is in the required task set $\{\texttt{Task}(m) \rightarrow x_m\}_{m=1}^n$.

## 2.1.2  Secure Hash Function

The secure hash function can be transformed into a fixed length output according to the input of any length: $h_s : \{0, 1\}^* \rightarrow Z_p^*$. It has the following two characteristics:

- Unidirectionality: given $x$, the calculation of $h_s(x)$ is simple. Given $h_s(x)$, the calculation of $x$ is very difficult.
- Collision Resistance: given $x = y$, it is simple to get $h_s(x) = h_s(y)$. It is very difficult to find two different values that $x \neq y$ to make $h_s(x) = h_s(y)$.

In the privacy-preserving content-based task allocation designed in this book, the secure hash function is used to transform task information into hash value and generate the task dictionary.

## 2.1.3  Searchable Encryption

Searchable encryption (SE) enables the third-party platform to find data or users that meet the requirements of data requesters in the form of ciphertext. From the perspective of encryption algorithms used, searchable encryption can be roughly divided into symmetric searchable encryption and public-key searchable encryption. This part mainly introduces the framework of searchable encryption scheme. On the whole, the searchable encryption scheme includes three entities: third-party platform, data holder, and data requester. Its basic process is summarized as follows:

- System Initialization: the data holder generates the key $sk$ and shares it with the data requester.
- Data Encryption: the data holder encrypts the data $w$ with the key and sends the ciphertext $E[w]$ to the third-party platform.
- Trapdoor Generation: for the search keyword $w'$, the data requester uses the key to generate $T[w']$ and sends it to the third-party platform.
- Data Retrieval: the third-party platform calculates the ciphertext $E[w]$ and trapdoor $T[w']$. If $w = w'$, output 1; otherwise, output 0.

If the SE scheme outputs 1 for any $w = w'$, then the SE scheme is correct. This book uses the SE framework to design the privacy-preserving task allocation.

### 2.1.4 Asymmetric Scalar-Product-Preserving Encryption

The asymmetric scalar-product-preserving encryption (ASPE) was proposed by Wong et al. [1]. The algorithm can be used to construct a secure kNN algorithm, which can make the third-party platform obtain the inner product of $\mathbf{x}$ and $\mathbf{y}$ without knowing the plaintext of vector $\mathbf{x}$ and $\mathbf{y}$. That is, assume that $E[\mathbf{x}]$, $E[\mathbf{y}]$ are the ciphertext of $\mathbf{x}$ and $\mathbf{y}$, respectively, and $E[\mathbf{x}] \cdot E[\mathbf{y}] = \mathbf{x} \cdot \mathbf{y}^T$. The specific flow of secure kNN algorithm is as follows:

- Initialization: the system randomly generates two $n \times n$ invertible matrices $\{M_1, M_2\}$ and an n-dimensional random vector $S \in \{0, 1\}^n$.
- Data Encryption: given the n-dimensional vector $\mathbf{x} = (x_1, x_2, \ldots, x_n)$, divide $\mathbf{x}$ into two random vectors $\mathbf{x_1}$ and $\mathbf{x_2}$. Specifically, for the $k(k \in [1, n])$ bit of $\mathbf{x}$, if $S(k) = 0$, set $\mathbf{x_1}(k) = \mathbf{x_2}(k) = x_k$; if $S(k) = 1$, set $\mathbf{x_1}$ and $\mathbf{x_2}$ to randoms such that $\mathbf{x_1}(k) + \mathbf{x_2}(k) = x_k$. Then, the user encrypts $\mathbf{x}$ as $(\mathbf{x_1} \times M_1, \mathbf{x_2} \times M_2)$.
- Trapdoor Generation: given the n-dimensional vector $\mathbf{y} = (y_1, y_2, \ldots, y_n)$, divide $\mathbf{y}$ into two random vectors $\mathbf{y_1}$ and $\mathbf{y_2}$. Specifically, for the $k(k \in [1, n])$ bit of $\mathbf{y}$, if $S(k) = 0$, set $\mathbf{y_1}$ and $\mathbf{y_2}$ to randoms such that $\mathbf{y_1}(k) + \mathbf{y_2}(k) = y_k$; if $S(k) = 1$, set $\mathbf{y_1}(k) = \mathbf{y_2}(k) = y_k$. Then, the user encrypts $\mathbf{y}$ as $(M_1^{-1} \times \mathbf{y_1^T}, M_2^{-1} \times \mathbf{y_2^T})$.
- Data Matching: the third-party platform calculates

$$
\begin{aligned}
\Delta &= (\mathbf{x_1} \times M_1, \mathbf{x_2} \times M_2) \cdot \left(M_1^{-1} \times \mathbf{y}_1^T, M_2^{-1} \times \mathbf{y}_2^T\right) \\
&= \mathbf{x}_1 \cdot \mathbf{y}_1^T + \mathbf{x}_2 \cdot \mathbf{y}_2^T = \mathbf{x} \cdot \mathbf{y}^T.
\end{aligned}
\tag{2.3}
$$

The secure kNN algorithm is mostly used in searchable encryption schemes and has high requirements for security model. In order to achieve higher security, Zhou et al. [2] designed another asymmetric scalar product calculation encryption algorithm: threshold predicate encryption (TPE). The specific flow of the algorithm is as follows:

- Initialization: the user randomly generates two $(n+3)\times(n+3)$ invertible matrices $\{M_1, M_2\}$, shuffle function $\pi : \mathbb{R}^{n+3} \to \mathbb{R}^{n+3}$, and threshold $\theta$.
- Data Encryption: the user randomly generates real numbers $\beta > 0, r_x$. Given the n-dimensional vector $\mathbf{x} = (x_1, x_2, \ldots, x_n)$, the user extends it to the $(n + 3)$-dimensional $\mathbf{x}' = (\beta x_1, \beta x_2, \ldots, \beta x_n, -\beta \theta, r_x, 0)$, perturbs $\mathbf{x}'$ into $\mathbf{x}'' = \pi(\mathbf{x}')$, expands $\mathbf{x}''$ into the lower triangular random matrix $S_x$ whose diagonal element is $\mathbf{x}''$, and encrypts $S_x$ into

$$
C_x = M_1 \times S_x \times M_2.
\tag{2.4}
$$

- Trapdoor Generation: the user randomly generates real numbers $\alpha > 0, r_y$. Given the n-dimensional vector $\mathbf{y} = (y_1, y_2, \ldots, y_n)$, the user extends it to the $(n + 3)$-dimensional $\mathbf{y}' = (\alpha y_1, \alpha y_2, \ldots, \alpha y_n, -\alpha, 0, r_y)$, perturbs $\mathbf{y}'$ into $\mathbf{y}'' = \pi(\mathbf{y}')$, expands $\mathbf{y}''$ into the lower triangular random matrix $S_y$ whose diagonal element is $\mathbf{y}''$, and encrypts $S_y$ into

$$C_y = M_2^{-1} \times S_y \times M_1^{-1}. \tag{2.5}$$

- Data Matching: the third-party platform calculates

$$\Delta = \mathrm{tr}\left(C_x \times C_y\right) = \alpha\beta \left(\mathbf{x} \cdot \mathbf{y}^T - \theta\right), \tag{2.6}$$

where $tr(\cdot)$ is the operation of matrix trace. If $\Delta \leq 0$, return 1; otherwise, 0 is returned.

TPE algorithm realizes the security under selective plaintext attack for external attackers. However, the algorithm only supports the matching of users' own data, so it can only be applied to limited scenes, such as fingerprint identification. Based on the secure kNN algorithm and TPE algorithm, this book designs a new secure and efficient asymmetric scalar-product-preserving encryption algorithm which can be applied to MCS scene.

## 2.1.5 Polynomial Fitting

The polynomial fitting technology can fit the curve according to the coordinates of multiple points. As shown in Fig. 2.1, for the geometric range $(\theta_a, \theta_b)$, select several coordinate points on the curve $(\theta_a, \theta_b)$ to fit two curves $(\theta_a^*, \theta_b^*)$, so that the generated curve is consistent with the original geometric curve as much as possible. The fitted curve equation can be expressed as $\theta_a^*(x) = a_0 + a_1 x + a_2 x^2 + \cdots + a_n x^n$,

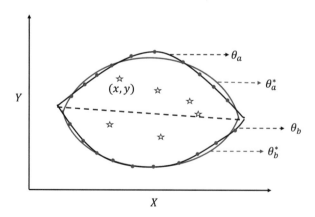

**Fig. 2.1** Polynomial fitting

$\theta_b^*(x) = b_0 + b_1 x + b_2 x^2 + \cdots + b_n x^n$, where $a_i$, $b_i (i \in [0, n])$ are the coefficients of the two curve equations, respectively, and $n$ is the highest power of the curve equation. Given geometric range $(\theta_a^*, \theta_b^*)$, judge whether a node $(x_i, y_i)$ is within the geometric range through the following steps:

- Step 1: calculate $(\theta_a^*(x_i) - y_i)$ and judge whether the value is greater than 0. If it is greater than 0, proceed to step 2; otherwise, the node is not within the geometric range.
- Step 2: calculate $(\theta_b^*(x_i) - y_i)$ and judge whether the value is less than 0. If it is less than 0, the node is within the geometric range; otherwise, the node is not within the geometric range.

The polynomial fitting technology is an approximate algorithm. The fitting function generated by this technology may not fit the given geometric range perfectly, resulting in certain fitting error. In order to improve the fitting accuracy, the fitting error can be controlled in a very small range through techniques such as orthogonal family. Interested readers can refer to papers [3, 4] for more technical details. In this book, the polynomial fitting technology is used to fit any geographical range and, based on the fitted curve, judge whether the location of the perceived user is within the geographical range queried by the data requester.

## 2.2  Preliminary of Privacy-Preserving Data Analysis Techniques

### 2.2.1  Truth Discovery

A truth discovery algorithm is an effective technology to calculate the real results of each task from multi-task data. On the whole, truth discovery mainly includes two steps: weight update and truth update. Suppose there are $K$ users and $M$ perception tasks, the perception data of user $u_k$ for task $o_m$ is $x_m^k$, the weight of user $u_k$ is $w_k$, and the truth of task $o_m$ is $x_m^*$. The following two steps are iterated until the iteration ends, and the truth of each task can be obtained:

- Weight Update: this step calculates the weight of the user according to the user data and the task truth. Specifically, the user's weight can be calculated as

$$w_k = f \left( \sum_{m=1}^{M} d\left(x_m^k, x_m^*\right) \right),$$ 

(2.7)

where $f(\cdot)$ is a monotonically decreasing function and $d(x^k{}_m, x^*{}_m)$ represents the distance between user data and truth.

- Truth Update: this step calculates the truth of the task according to the user's weight and data. Specifically, the truth can be calculated as

$$x_m^* = \frac{\sum_{k=1}^K w_k \cdot x_m^k}{\sum_{k=1}^K w_k}.$$

(2.8)

In weight updating, the distance calculation function $d(\cdot)$ is slightly different according to different data types. For continuous data (such as speed, temperature, humidity, etc.), the square distance function $d(x^k{}_m, x_m^*) = (x_m^k - x_m^*)^2$ is used to represent the distance between user data and truth; for discrete data (such as question and answer options), assume that task $o_m$ has several answer options, $x_m^k = (0, \ldots, \overset{q}{1}, \ldots, 0)^T$ represents the q-th answer selected by user $u_k$, and $d(x_m^k, x_m^*) = (x_m^k - x_m^*)^T (x_m^k - x_m^*)$ represents the distance between user data and truth. Consistent with most of the existing truth discovery schemes [5, 6], this book selects the logarithmic equation to calculate the user's weight, i.e.,

$$w_k = \log\left(\frac{\sum_{k=1}^K \sum_{m=1}^M d\left(x_m^k, x_m^*\right)}{\sum_{m=1}^M d\left(x_m^k, x_m^*\right)}\right).$$

(2.9)

In the privacy protection truth discovery scheme of truth disclosure and truth hiding designed in this book, the truth discovery algorithm is used to calculate the truth of the task.

## 2.2.2   Public-Key Cryptosystem Supporting Distributed Decryption

Liu et al. [7] proposed the public-key cryptosystem supporting distributed decryption (PCDD) based on the public-key encryption algorithm [8]. PCDD splits the decrypted private key and distributes it to several entities. Multiple entities decrypt ciphertext through cooperation. Because of its high efficiency and scalability, PCDD is widely used in secure multi-party computing and other applications. It mainly includes the following algorithms:

- **KeyGen**$(\kappa)$(Initialization): given a security parameter $\kappa$ select two large prime numbers $p, q, |p| = |q| = \kappa$, and calculate $n = pq, \lambda = lcm(p-1)(q-1)/2$, where $lcm(a, b)$ is the least common multiple of $a$ and $b$. Select random number $g \in Z_{n^2}^*$ and random number $\theta \in [1, n/4]$, and calculate $h = g^\theta \mod n^2$. The public key is $pk = (n, g, h)$, and the weak private key is $\theta$, and the strong private key is $\lambda$.
- **Enc**$(pk, m)$(Data Encryption): given the data $m \in Z_n$, the algorithm encrypts $m$ as $\mathbf{E}[m] = h^r(1 + nm) \mod n^2, r \in [1, n/4]$.

- **SDec**$(\lambda, \mathbf{E}[m])$(Ciphertext Decryption): given the ciphertext $\mathbf{E}[m]$, the algorithm uses a strong private key to decrypt the ciphertext $m = L((\mathbf{E}[m])^\lambda \bmod n^2)\lambda^{-1} \bmod n$, where $(\mathbf{E}[m])^\lambda \bmod n^2 = g^{\theta r \lambda}(1+mn\lambda) \bmod n^2$, $L(x) = \frac{x-1}{n}$.
- **SkeyS**$(\lambda, t)$(Private Key Decomposition): given the strong private key $\lambda$, the algorithm divides $\lambda$ into t parts $(\lambda_1, \lambda_2, \ldots, \lambda_t)$, where $\sum_{i=1}^{t} \lambda_i \equiv 0 \bmod \lambda$, $\sum_{i=1}^{t} \lambda_i \equiv 1 \bmod n^2$.
- **PSDec**$(\lambda, \mathbf{E}[m])$(Partial Decryption): given the ciphertext $\mathbf{E}[m]$, the algorithm partially decrypts the ciphertext as $CT^{(i)} = (\mathbf{E}[m])^\lambda = g^{\theta r \lambda_i}(1+mn\lambda_i) \bmod n^2$.
- **DDec**$(\{CT^{(i)}_{i=1}^{t}\})$(Complete Decryption): given the partially decrypted ciphertext $CT^{(1)}, CT^{(2)}, \ldots, CT^{(t)}$, the algorithm performs complete decryption $m = L(\prod_{i=1}^{t} CT^{(i)})$.

For any $m \in Z_n$, $(\mathbf{E}[m])^{n-1} = h^{r(n-1)} \cdot (1 + (n-1)mn) \bmod n^2 = \mathbf{E}[-m]$. For the convenience of narration, this book uses $(\mathbf{E}[m])^{-1}$ to represent $(\mathbf{E}[m])^{n-1}$. PCDD has a homomorphic property similar to Paillier encryption system, that is, given plaintext $m_1, m_2, a \in Z_n$, PCDD satisfies

$$E[m_1] \cdot E[m_2] = E[m_1 + m_2]$$
$$(E[m])^a = E[a \cdot m]. \tag{2.10}$$

In the privacy protection truth discovery scheme of truth disclosure and truth hiding designed in this book, PCDD is used to protect data privacy and construct privacy protection truth discovery algorithm.

# References

1. Wong, W.K., Cheung, D.W.-l., Kao, B., Mamoulis, N.: Secure kNN computation on encrypted databases. In: Proceedings of the 2009 ACM SIGMOD International Conference on Management of Data, pp. 139–152 (2009)
2. Zhou, K., Ren, J.: Passbio: privacy-preserving user-centric biometric authentication. IEEE Trans. Inf. Forensics Secur. **13**(12), 3050–3063 (2018)
3. Xu, G., Li, H., Dai, Y., Yang, K., Lin, X.: Enabling efficient and geometric range query with access control over encrypted spatial data. IEEE Trans. Inf. Forensics Secur. **14**(4), 870–885 (2018)
4. Strobach, P.: Solving cubics by polynomial fitting. J. Comput. Appl. Math. **235**(9), 3033–3052 (2011)
5. Li, Q., Li, Y., Gao, J., Zhao, B,, Fan, W., Han, J.: Resolving conflicts in heterogeneous data by truth discovery and source reliability estimation. In: Proceedings of the 2014 ACM SIGMOD International Conference on Management of Data, pp. 1187–1198 (2014)
6. Li, Y., Li, Q., Gao, J., Su, L., Zhao, B., Fan, W., Han, J.: Conflicts to harmony: a framework for resolving conflicts in heterogeneous data by truth discovery. IEEE Trans. Knowl. Data Eng. **28**(8), 1986–1999 (2016)

7. Liu, X., Qin, B., Deng, R.H., Lu, R., Ma, J.: A privacy-preserving outsourced functional computation framework across large-scale multiple encrypted domains. IEEE Trans. Comput. **65**(12), 3567–3579 (2016)
8. Bresson, E., Catalano, D., Pointcheval, D.: A simple public-key cryptosystem with a double trapdoor decryption mechanism and its applications. In: International Conference on the Theory and Application of Cryptology and Information Security, pp. 37–54 (2003)

# Part II
# Privacy-Preserving Task Allocation

# Chapter 3
# Privacy-Preserving Content-Based Task Allocation

**Abstract** Content-based task allocation is an important part of MCS, which can effectively help data requesters find suitable sensing users to collect data. However, users' task information may reveal sensitive information, e.g., location and hobbies. An intuitive scheme is to encrypt task information and perform task queries in ciphertexts. However, considering that MCS has the characteristics of a large scale of users, many types of tasks, and resource-limited devices, schemes based on traditional time-consuming cryptographic tools will consume tremendous computing resources of the mobile crowdsensing platform. In order to solve the challenges, this chapter firstly proposes a basic privacy-preserving content-based task allocation scheme, named PPTA, based on random matrix multiplication technology and polynomial functions, which can achieve efficient task queries without revealing task privacy. In addition, this chapter extends PPTA so that it can efficiently support more functionalities in task allocation, such as conjunctive task allocation, Top-z task allocation, task allocation with access control, and task recovery. Through detailed security analysis, we demonstrate that PPTA can well preserve task privacy and query privacy. Experiments based on real mobile crowdsensing applications demonstrate that PPTA can achieve high data encryption and task query efficiency.

## 3.1 Introduction

This section describes the overview, related works, and preliminary of this chapter.

### 3.1.1 Overview

With the development of 5G/6G communication technology [1–4] and the widespread popularity of various sensing sensors [5–7], MCS has become an effective way of data sensing and collection [8–11]. At present, there are many commercial mobile crowdsensing platforms that can establish connections between

© The Author(s), under exclusive license to Springer Nature Singapore Pte Ltd. 2023   33
C. Zhang et al., *Privacy-Preserving in Mobile Crowdsensing*,
https://doi.org/10.1007/978-981-19-8315-3_3

data requesters and sensing users, such as Amazon Mechanical Turk,[1] WAZE,[2] Be My Eyes,[3] Witmart,[4] Zhihu,[5] etc., which have promoted a large number of applications such as health data collection [12], public opinion collection [13], and text translation [14]. Since there are usually a large number of users and tasks in the mobile crowdsensing platform, in order to ensure the efficiency of data collection and accuracy of service, it is particularly important for the mobile crowdsensing platform to allocate different tasks to appropriate users. In recent years, the problem of "task allocation" in MCS has attracted extensive attention from academia and industry.

In the existing task allocation schemes, each data requester sends its own task requirements (usually expressed by task keywords) to the mobile crowdsensing platform, and then the mobile crowdsensing platform retrieves its own database to find the sensing users who meet the data requester's task requirements. Although this approach is effective, performing the query on task plaintexts may bring serious privacy leakage problems. In actual mobile crowdsensing applications, the task information uploaded by users is often closely related to sensitive information such as their identity, health status, location, and hobbies. For example, a driver's participation in the task of collecting traffic information can help other users to better understand the current traffic situation, but its own position, trajectory, and other information may be inferred from the task it participated in; a patient with a certain disease can help medical institutions evaluate the efficacy of new drugs by participating in new drug testing, but their task of participating in new drug testing may reveal their health or economic status. In addition, as a third-party platform, the mobile crowdsensing platform is usually not fully trusted, and it may leak the user's task information to other institutions for more benefits [15, 16]. If the task information is not effectively preserved, users may be reluctant to send their own data to the mobile crowdsensing platform, which greatly affects the healthy development of mobile crowdsensing applications.

An intuitive way to achieve privacy-preserving task allocation is that the user first encrypts the task information, and then the mobile crowdsensing platform performs the query in the form of ciphertext. Therefore, searchable encryption is a promising technology to achieve privacy-preserving task allocation based on task content. However, most of the current schemes based on searchable encryption technology [17–21] adopt the trusted data center model. That is, these schemes rely on a trusted data center to encrypt data and distribute search keys to search users. This implementation mode has some inherent flaws and is therefore not suitable for mobile crowdsensing scenarios. Firstly, sensing users need to upload their own data to the trusted data center, and then the trusted data center encrypts it and uploads it

---

[1] https://www.mturk.com/mturk/welcome.

[2] https://play.google.com/store/apps/details?id=com.waze.

[3] https://www.bemyeyes.com/.

[4] http://www.witmart.com/cn/.

[5] https://www.zhihu.com/hot.

to the mobile crowdsensing platform. This form of data encryption and transmission reduces system performance and increases the risk of data leakage. Secondly, because the search users all have the same search key, some users who hold the search key may recover the plaintext from the search ciphertext of other users. There are also some public-key searchable encryption schemes currently which can realize the privacy-preserving content-based task allocation without introducing a trusted data center. However, these schemes either fail to achieve effective privacy preservation or fail to achieve a high query efficiency. Among them, Tang et al. [22] designed a privacy-preserving task allocation scheme using bilinear mapping. However, due to the time-consuming pair operation of bilinear mapping, this scheme consumes a lot of computational resources of the mobile crowdsensing platform. Shu et al. [23] made appropriate improvements to the secure kNN technology [24], and based on this, they designed an efficient privacy-preserving task matching scheme. However, this scheme assumes that all users are completely trusted, and it is difficult to resist internal or external attacks in real mobile crowdsensing scenarios. For example, Gu et al. [25] proved that when the attacker can obtain some plaintext-ciphertext pair information, the key information in the secure kNN algorithm can be recovered. Ni et al. [26] designed a privacy-preserving task allocation scheme by exploiting the properties of matrices. The scheme uses a random matrix to hide the user's task information and ensures that the mobile crowdsensing platform performs tasks on the random matrix without knowing the user's task information. Compared with [23], Ni et al. [26] achieve a higher level of security by assuming that users are semi-trusted and achieve security under chosen-plaintext attacks. However, since the dimension of the matrix in this scheme is linearly related to the number of task types, when there are many types of tasks, this scheme will face a high computational cost.

In order to solve the above problems, this chapter studies the problem of privacy-preserving task allocation in mobile crowdsensing scenarios and proposes a privacy-preserving content-based task allocation scheme (PPTA). The main contributions of this chapter are summarized as follows:

- Identify the security and privacy issues of MCS privacy preservation task allocation and define the security model of task allocation in MCS according to the different attack capabilities of attackers.
- A privacy-preserving content-based task allocation scheme (PPTA) is proposed using random matrix multiplication technologies and polynomial functions. Specifically, PPTA utilizes the polynomial function to represent the user's task information, utilizes the random matrix to preserve the task information, and utilizes the properties of the matrix to perform query operations on the random matrix.
- Extend PPTA to further support more functions, such as conjunctive task allocation, Top-z task allocation, task allocation with access control, and task recovery, ensuring the versatility and practicability of PPTA.
- Security analysis proves that PPTA can effectively preserve users' task privacy and query privacy. Experiments based on real mobile crowdsensing scenarios prove that PPTA has high computational efficiency in the data encryption and query phases.

The rest of this chapter is organized as follows: Sect. 3.1.2 presents the related works on task allocation in MCS; Sect. 3.1.3 provides the brief description about polynomial function; Sect. 3.2 describes the architecture overview of PPTA from system model, security model, and design goals; Sect. 3.3 presents the detailed design of PPTA; Sect. 3.4 gives the security analysis of PPTA; Sect. 3.5 presents the performance evaluation and analysis; and Sect. 3.6 summarizes the work of this chapter.

## 3.1.2  Related Works

In this section, we briefly review existing related works on task allocation in MCS.

As an important part of MCS, task allocation has attracted extensive attention of many researchers in recent years. To help data requesters find suitable sensing users, researchers have designed a series of task allocation schemes. Among them, Yuen et al. [27] proposed a task allocation scheme in dynamic crowdsensing scenarios based on unified probability matrix decomposition. Yang et al. [28] considered the possibility of user participation in the task and calculated task recommendation scores for different users based on this. Karaliopoulos et al. [29] used logistic regression to calculate the likelihood of a user accepting a task. Kobren et al. [30] improved the accuracy of task allocation by considering the duration of user engagement in the task and the user's interest. Wang et al. [31] considered the spatiotemporal properties in MCS and proposed a pricing strategy to motivate users to participate in tasks. Tong et al. [32] formulated an efficient task allocation scheme using a greedy strategy.

As security and privacy issues become more and more important, in recent years, many researchers have devoted themselves to researching the privacy preservation of task allocation in MCS. Specifically, Gong et al. [33] proposed the first user privacy-preserving task allocation scheme, which formally transformed task recommendation into an optimization problem of privacy, practicality, and efficiency, and proposed a greedy algorithm to achieve a near-optimal scheme. However, this scheme does not strictly preserve the user's task privacy and may reveal sensitive information such as the user's identity, location, and hobbies. In order to realize the privacy preservation of task information and realize task allocation based on task content, Shu et al. proposed several task allocation schemes to preserve task privacy [23, 34, 35] to solve the problems of single-task matching, multi-task matching, witch attacks, etc. In particular, their scheme leveraged the secure kNN technology [23] to achieve high computational efficiency compared to existing schemes. Specifically, the scheme uses matrix decomposition to split the keys of the secure kNN algorithm, uses polynomial functions to generate task indices and query tokens, and uses random matrices to encrypt task information. Although this scheme can achieve high computational efficiency, it assumes that all users are completely trusted, which is a strong security assumption in real mobile crowdsensing scenarios. In fact, some users may obtain other users' private data

**Table 3.1** Comparison between PPTA and other works in terms of security and efficiency

| Scheme | Security | Computational efficiency |
|--------|----------|--------------------------|
| [22]   | IND-CPA  | Low                      |
| [23]   | IND-KPA  | Very high                |
| [26]   | IND-CPA  | High                     |
| [38]   | IND-CPA  | Low                      |
| PPTA   | IND-CPA  | Very high                |

based on the existing key and other information. To achieve more secure task allocation, Ni et al. designed a privacy-preserving task matching scheme based on random matrix multiplication technology [26, 36]. Compared with scheme [23], their scheme is resistant to chosen-plaintext attacks. Compared with schemes [22, 37], their scheme has higher computational efficiency. However, in this scheme, the matrix dimension increases with the number of task types, which makes the scheme still face large computational and communication overhead when there are many types of tasks. In addition, there are now some privacy-preserving task allocation schemes that consider other factors, such as considering user attributes [38], user capabilities [39], location privacy [40, 41], fog computing environments [42], and edge computing environments [43]. Table 3.1 lists the comparison between PPTA and existing related works. It can be seen that PPTA can achieve high security and high computational efficiency at the same time.

## 3.1.3 Preliminary

In this section, we give a brief description of the polynomial function before presenting the details of our proposed PPTA.

### 3.1.3.1 Polynomial Function

Given a polynomial function $f(x) = a_n x^n + a_{n-1} x^{n-1} + \cdots + a_1 x + a_0$ with the highest power of $n$, where the coefficient of $x^i$ is $a_i$, $i \in [0, n]$, and $a_n \neq 0$, according to the basic theorem of algebra,

$$
\begin{aligned}
f(x) &= a_n x^n + a_{n-1} x^{n-1} + \cdots + a_1 x + a_0 \\
&= a_n (x - x_n)(x - x_{n-1}) \cdots (x - x_1),
\end{aligned}
\tag{3.1}
$$

where $(x_1, x_2, \ldots, x_n)$ is the root of polynomial function $f(x)$. According to Weida's theorem, the $n - k$th coefficient of $f(x)$, i.e., $a_{n-k}(k \in [0, n])$, can be calculated as

$$\sum_{1 \le i_1 < i_2 < \cdots < i_k \le n} \left( \prod_{j=1}^{k} x_{i_j} \right) = (-1)^k \frac{a_{n-k}}{a_n}. \tag{3.2}$$

In PPTA, a polynomial function is used to judge whether a task $x_i$ belongs to the task set $\{x_1, x_2, \ldots, x_n\}$. That is, suppose $\{\mathtt{Task}(m) \to x_m\}_{m=1}^{n}$ is the required task set, and $f(x) = a_n(x - x_n)(x - x_{n-1}) \cdots (x - x_1)$ is the corresponding polynomial function. If $f(x_t) = 0$, the task $\mathtt{Task}(m) \to x_t$ is in the required task set $\{\mathtt{Task}(m) \to x_m\}_{m=1}^{n}$.

## 3.2 Architecture Overview

In this section, we will describe the system model, security model, and design goals of PPTA.

### 3.2.1 System Model

As shown in Fig. 3.1, PPTA contains four entities, namely, trusted authority (TA), mobile crowdsensing platform (CP), data requester (DR), and sensing user (SU).

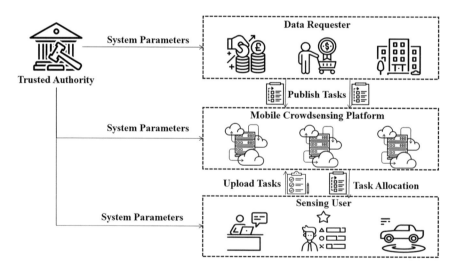

**Fig. 3.1**  System model of PPTA

The functions of each entity are defined as follows:

- Trusted authority: The trusted authority is responsible for generating relevant keys for entities participating in task allocation. After the system is initialized, the trusted authority will be offline or dormant.
- Mobile crowdsensing platform: The mobile crowdsensing platform is usually a third-party platform, such as a company or organization, which is responsible for establishing connections between data requesters and sensing users. It publishes tasks from data requesters and recruits sensing users who meet their task requirements.
- Data requester: The data requester has a series of task requirements. Since their own capabilities, data collection costs, etc., are limited, the data requester uploads tasks to the mobile crowdsensing platform, hoping that the mobile crowdsensing platform can find sensing users who meet their task requirements.
- Sensing user: Sensing users are typically individuals holding mobile devices. After accepting the task from the mobile crowdsensing platform, the sensing users utilize GPS, camera, accelerometer, temperature sensor, etc., in the mobile device to collect the corresponding data and upload it to the mobile crowdsensing platform.

## 3.2.2 Security Model

In PPTA, we assume that the trusted authority is fully trusted and that the communication channel between the trusted authority and any other entity is secure. After the system is initialized, the trusted authority can remain offline. The other entities are honest-but-curious. Specifically, these entities will honestly perform the designed protocol, but at the same time, they will do everything possible to obtain other users' private information. Additionally, as with most existing task allocation schemes, we assume that the mobile crowdsensing platform does not collude with and among users [22, 23]. This assumption is reasonable because (1) mobile crowdsensing platforms are usually run by large companies or institutions that fully understand the importance of reputation and (2) in order to preserve their privacy, users will not share their own key information with other users. Based on the different attack capabilities of the adversary, we define the following two attacks:

- Passive attack: Passive attack means that the adversary can only obtain the ciphertext information. This attack is similar to a ciphertext-only attack in cryptography. This attack may be launched when an adversary monitors a user's communication channel or the mobile crowdsensing platform receives the user's ciphertext data.
- Active attack: In an active attack, the adversary may obtain more information in addition to the ciphertext information. We give specific active attack modes from the perspective of internal attackers of mobile crowdsensing platforms and non-mobile crowdsensing platforms. For internal attackers who are in the system,

they may obtain the correspondence between some known tasks and ciphertexts. This attack is similar to the known-plaintext attack in cryptography. For internal attackers who are not in the system, they may attempt to obtain sensitive information of other users based on information such as task dictionaries and distribution keys. For instance, if a user's key is stored in a physical device (such as a smart card) [23], once the adversary (such as a user with encryption keys) obtains the users' smart cards, the adversary may choose special plaintext as input and observe the output ciphertexts, and then try to obtain the key information in the smart card based on these plaintext-ciphertext pairs. This attack is similar to the chosen-plaintext attack in cryptography.

Note that the mobile crowdsensing platform may also obtain information, such as the query frequency of task information, and then infer the user's private information. This attack is similar to the hot-word attack [44] or the known background attack [45]. For this kind of attack, there are already some good schemes, such as group-based keyword construction mode [44]. Specifically, these schemes usually randomly group task keywords to defend against these attacks and are applicable to the majority of existing task allocation schemes. Therefore, we do not consider such attacks in this chapter. Readers are referred to the paper [44] for more technical details.

### 3.2.3   Design Goals

Suppose there are now $M$ mobile crowdsensing tasks and $K$ sensing users. Suppose $T_{k,m}$ is the $m$-th task that sensing user $u_k$ is interested in, and $\mathbb{T}_J$ represents the set of tasks submitted by data requester $u_j$. In PPTA, the mobile crowdsensing platform finds the sensing users who meet the task requirements of the data requester, namely, $T_{k,m} \in \mathbb{T}_j$, and at the same time, the query of the data requester and the task information of the sensing user are not compromised in the given security model. Specifically, PPTA should achieve the following privacy and functionality goals:

- Data privacy: PPTA should effectively preserve users' task privacy and query privacy. That is, given a ciphertext, the adversary cannot recover the user's task or query information.
- Query accuracy: PPTA should guarantee query accuracy. That is, given a query, PPTA should accurately find sensing users that meet the data requester's task requirements.
- Query efficiency: PPTA should enable efficient query. That is, given a query, PPTA should perform ciphertext query operations efficiently.

## 3.3 Detailed Design

In this section, we give the detailed content of PPTA. Specifically, we give the detailed design of PPTA, analyze its correctness, and then extend PPTA to support more functional requirements in MCS.

### 3.3.1 Proposed PPTA Scheme

**System Initialization** The trusted authority generates encryption keys and re-encryption keys for the other entities in the system. Specifically, given the highest power of polynomial function $n$, the trusted authority generates six random $(n + 2) \times (n+2)$-dimensional invertible matrices $\{M_1, M_2, A_1, A_2, B_1, B_2\}$ as the master secret key. Then, for each registered sensing user $u_k$, the trusted authority chooses a random $(n + 2) \times (n + 2)$-dimensional lower triangular matrix $I_k$ with main diagonal elements $(1, 1, 1, \cdots, 1, 0)$ and generates encryption key $A_{k,1}$ and $A_{k,2}$, where $A_{k,1} = A_1^{-1} \times M_1 \times I_k$ and $A_{k,2} = I_k \times M_2 \times A_2^{-1}$. For each registered data requester $u_j$, the trusted authority chooses a random $(n + 2) \times (n + 2)$-dimensional lower triangular matrix $Q_j$ with main diagonal elements $(1, 1, 1, \cdots, 1, 0)$ and generates encryption key $B_{j,1}$ and $B_{j,2}$, where $B_{j,1} = B_1^{-1} \times M_2^{-1} \times Q_j$ and $B_{j,2} = Q_j \times M_1^{-1} \times B_2^{-1}$. Next, the trusted authority distributes the encryption key to the corresponding user and distributes the re-encryption key $\{A_1, A_2, B_1, B_2\}$ to the mobile crowdsensing platform through a secure channel.

In addition, the trusted authority generates a task dictionary $Task$, which maps different tasks to different random numbers. In practical applications, a secure hash function can be used to map tasks to random numbers, which leverages the one-way and collision resistance to preserve the privacy of tasks. Then, the trusted authority sends the task dictionary $Task$ to sensing users and data requesters through a secure communication channel. After completing the system initialization, the trusted authority can be offline or dormant.

**Task Encryption** The sensing user $u_k$ specifies the tasks they are interested in, encrypts the tasks, and sends the encrypted tasks to the mobile crowdsensing platform. Specifically, given a set of tasks $\{Task(m)\}_{m=1}^{l}$, $u_k$ obtains the corresponding data $\{Task_{k,m}\}_{m=1}^{l}$ from the task dictionary $Task$. To keep the dimensions consistent, if $l < n$, $u_k$ additionally selects some random numbers $\{Task_{k,m}\}_{m=l+1}^{n}$, where $\{Task_{k,m}\}_{m=l+1}^{n} \notin \{Task(m)\}_{m=1}^{l}$. Then, $u_k$ chooses a random $a_n \neq 0$ and generates a polynomial function $f_{task_k}(x)$ to hide the task information they are interested in as follows:

$$f_{task_k}(x) = a_n(x - T_{k,1})(x - T_k, 2)\cdots(x - T_{k,n})$$

$$= a_0 + a_1 x + a_2 x^2 + \cdots + a_n x^n.$$

(3.3)

Next, $u_k$ extracts the coefficients $(a_0, a_1, a_2, \ldots, a_n)$ from the above polynomial function, chooses a random $r_k \neq 0$, and then generates a $(n + 2)(n + 2)$-dimensional lower triangular random matrix $\tau_k$ with main diagonal elements $(r_k a_0, r_k a_1, r_k a_2, \ldots, r_k a_n, 0)$ as follows:

$$
\tau_k = \begin{bmatrix}
r_k a_0 & 0 & 0 & 0 & \cdots & 0 & 0 & 0 \\
* & r_k a_1 & 0 & 0 & \cdots & 0 & 0 & 0 \\
\cdots & \cdots & \cdots & \cdots & & \cdots & \cdots & 0 \\
* & \cdots & * & r_k a_{i-1} & 0 & \cdots & 0 & 0 \\
* & \cdots & \cdots & * & r_k a_i & \cdots & 0 & 0 \\
\cdots & \cdots & \cdots & \cdots & & \cdots & \cdots & 0 \\
* & \cdots & \cdots & \cdots & \cdots & * & r_k a_n & 0 \\
* & \cdots & \cdots & \cdots & \cdots & * & * & 0
\end{bmatrix}_{(n+2) \times (n+2)},
$$

where $*$ represents the random in the lower triangular random matrix. Then, $u_k$ utilizes the encryption key $\{A_{k,1}, A_{k,2}\}$ to encrypt the $\tau_k$ as follows:

$$
E[\tau_k] = A_{k,1} \times \tau_k \times A_{k,2}. \tag{3.4}
$$

At last, $u_k$ uploads the ciphertext $E[\tau_k]$ to the mobile crowdsensing platform.

**Task Transformation**  After receiving the encrypted task data $E[\tau_k]$, the mobile crowdsensing platform utilizes the re-encryption key $\{A_1, A_2\}$ to perform task transformation. Specifically, the calculation is as follows:

$$
\begin{aligned}
RE[\tau_k] &= A_1 \times E[\tau_k] \times A_2 \\
&= A_1 \times A_{k,1} \times \tau_k \times A_{k,2} \times A_2 \\
&= M_1 \times I_k \times \tau_k \times I_k \times M_2.
\end{aligned} \tag{3.5}
$$

**Trapdoor Generation**  The data requester $u_j$ generates the trapdoor for the task that $u_j$ is interested in. Specifically, $u_j$ first specifies the task keywords $T_{j,m} \leftarrow Task(m)$ from the task dictionary $Task$. Then, for each $T_{j,m}$, $u_j$ chooses a random $r_{j,m} \neq 0$ and generates a $(n + 2) \times (n + 2)$ lower triangular random matrix $\tau_{j,m}$ with main diagonal elements $(r_{j,m}, r_{j,m} T_{j,m}, r_{j,m} T_{j,m}^2, \ldots, r_{j,m} T_{j,m}^n, 0)$. Next, $u_j$ generates the trapdoor and sends it to the mobile crowdsensing platform. Specifically, the calculation is as follows:

$$
T[\tau_{j,m}] = B_{j,1} \times \tau_{j,m} \times B_{j,2}. \tag{3.6}
$$

**Trapdoor Transformation**  After receiving the trapdoor data $T[\tau_{j,m}]$, the mobile crowdsensing platform utilizes the re-encryption key $\{B_1, B_2\}$ to perform task transformation. Specifically, the calculation is as follows:

$$
\begin{aligned}
RT[\tau_{j,m}] &= B_1 \times T[\tau_{j,m}] \times B_2 \\
&= B_1 \times B_{j,1} \times \tau_{j,m} \times B_{j,2} \times B_2 \quad\quad (3.7) \\
&= M_2^{-1} \times Q_j \times \tau_{j,m} \times Q_j \times M_1^{-1}.
\end{aligned}
$$

**Query**  The mobile crowdsensing platform utilizes the re-encrypted trapdoor to perform query operations on the re-encrypted task data stored in its database. Specifically, for each task ciphertext $RE[\tau_k]$, the mobile crowdsensing platform computes

$$
\begin{aligned}
\Delta_{k,(j,m)} &= tr(RE[\tau_k] \times RT[\tau_{j,m}]) \\
&= tr(M_1 \times I_k \times \tau_k \times I_k \times M_2 \times M_2^{-1} \times Q_j \times \tau_{j,m} \times Q_j \times M_1^{-1}) \\
&= tr(M_1 \times I_k \times \tau_k \times I_k \times Q_j \times \tau_{j,m} \times Q_j \times M_1^{-1}) \\
&= r_k r_{j,m} f_{task_k}(T_{j,m}),
\end{aligned}
$$
$$(3.8)$$

where $tr(\cdot)$ represents the matrix trace operation. Since $r_k$ and $r_{j,m}$ do not equal to 0, if $\Delta_{k,(j,m)} = 0$, then $f_{task_k}(T_{j,m}) = 0$. That is, the sensing user is interested in the $T_{j,m}$. Then, the mobile crowdsensing platform establishes the connection between the data requester and the sensing user. The detailed steps are as shown in Fig. 3.2.

### 3.3.2  Correctness Analysis

In this section, we present the correct analysis of PPTA. According to the design details of PPTA, the lower triangular random matrix is used to hide the user's task information, and the matrix decomposition is used to construct task encryption and task matching algorithms. To facilitate the proof, we first give two lemmas related to the matrix.

**Lemma 3.1**  *Given an invertible matrix $M$ and a square matrix $P$ with the same dimension, $tr(M \times P \times M^{-1}) = tr(P)$.*

***Proof***  According to the definition of linear algebra, $P \rightarrow M \times P \times M^{-1}$ is called the similarity transformation. According to the properties of a similar matrix, it can be known that after the matrix undergoes similarity transformation, the trace of the matrix (the sum of the main diagonal elements) remains unchanged. That is, the matrix and its similar matrix have the same trace. Thus, Lemma 3.1 is proven.

**System Initialization:** Given the highest power $n$ of a polynomial fitting function, the trusted authority generates six $(n+2) \times (n+2)$-dimensional invertible random matrices $\{M_1, M_2, A_1, A_2, B_1, B_2\}$ as the master secret key.

For a sensing user $u_k$, the trusted authority generates the encryption key $A_{k,1}$ and $A_{k,2}$ as follows:

$$A_{k,1} = A_1^{-1} \times M_1 \times I_k,$$
$$A_{k,2} = I_k \times M_2 \times A_2^{-1}. \tag{3.9}$$

For a data requester $u_q$, the trusted authority generates the encryption key $(B_{j,1}$ and $B_{j,2})$ as follows:

$$B_{j,1} = B_1^{-1} \times M_2^{-1} \times Q_j,$$
$$B_{j,2} = Q_j \times M_1^{-1} \times B_2^{-1}. \tag{3.10}$$

**Task Encryption:** Given the task interest set $\{Task_{k,m}\}_{m=1}^l$ that the sensing user $u_k$ takes interest in, $u_k$ generates a polynomial function $f_{task_k}(x) = a_n(x - T_{k,1})(x - T_{k,2}) \cdots (x - T_{k,n}) = a_0 + a_1 x + \cdots + a_n x^n$, where $a_n \neq 0$. Then, $u_k$ chooses a random $r_k \neq 0$ and generates $(n+2)(n+2)$-dimensional lower triangular random matrix $\tau_k$ with main diagonal elements $(r_k a_0, r_k a_1, r_k a_2, \cdots, r_k a_n, 0)$. Subsequently, $u_k$ encrypts $\tau_k$ as follows:

$$E[\tau_k] = A_{k,1} \times \tau_k \times A_{k,2}. \tag{3.11}$$

**Task Transformation:** After receiving the encrypted task data $E[\tau_k]$, the mobile crowd-sensing platform utilizes the re-encryption key $\{A_1, A_2\}$ to perform task transformation as follows:

$$RE[\tau_k] = A_1 \times E[\tau_k] \times A_2. \tag{3.12}$$

**Trapdoor Generation:** Given a task interest $T_{(m)}$ that the data requester $u_j$ requires, $u_j$ generates task information $T_{j,m} \leftarrow T(m)$. Then, $u_j$ chooses a random $r_{j,m} \neq 0$ and generates a $(n+2) \times (n+2)$ lower triangular random matrix $\tau_{j,m}$ with main diagonal elements $(r_{j,m}, r_{j,m}T_{j,m}, r_{j,m}T_{j,m}^2, \cdots, r_{j,m}T_{j,m}^n, 0)$. Subsequently, $u_j$ generates the trapdoor as follows:

$$T[\tau_{j,m}] = B_{j,1} \times \tau_{j,m} \times B_{j,2}. \tag{3.13}$$

Finally, the data requester $u_j$ sends $T[\tau_{j,m}]$ to the mobile crowdsensing platform.

**Trapdoor Transformation.** After receiving $T[\tau_{j,m}]$, the mobile crowdsensing platform utilizes the re-encryption key $\{B_1, B_2\}$ to perform task transformation as follows:

$$RT[\tau_{j,m}] = B_1 \times T[\tau_{j,m}] \times B_2. \tag{3.14}$$

**Query.** Based on the transformed task interests and trapdoor, the mobile crowdsensing platform computes

$$\Delta_{k,(j,m)} = tr(RE[\tau_k] \times RT[\tau_{j,m}]). \tag{3.15}$$

If $\Delta_{k,(j,m)} = 0$, the sensing user is interested in the $T_{j,m}$. Otherwise, the sensing user is not interested in the $T_{j,m}$.

**Fig. 3.2** Detail in PPTA

**Lemma 3.2** *Given two lower triangular random matrices $P$ and $S$ with the same dimension, if their main diagonal elements are vectors $\mathbf{P}$ and $\mathbf{S}$, respectively, then $tr(P \times S) = \mathbf{P} \circ \mathbf{S}$, where $\mathbf{P} \circ \mathbf{S}$ represents the inner product of $\mathbf{P}$ and $\mathbf{S}$.*

**Proof** According to the definition of linear algebra, the matrix obtained by multiplying two lower triangular matrices is still a lower triangular matrix. According to the properties of matrix multiplication, it can be known that the main diagonal elements of the lower triangular matrix obtained by multiplying two lower triangular matrices are equal to the result of multiplying the corresponding main diagonal elements of the two lower triangular matrices. Thus, Lemma 3.2 is proven.

**Lemma 3.3** *Assuming that there exist polynomial function $f_{task_k}(x)$ and task $T_{j,m}$. The coefficients of $f_{task_k}(x)$ are expanded and encrypted as $RE[\tau_k]$, and the coefficients of $T_{j,m}$ are expanded and encrypted as $RT[T_{j,m}]$. When $f_{task_k}(x) = 0$, $\Delta_{k,(j,m)} = tr(RE[\tau_k] \times RT[\tau_{j,m}]) = 0$.*

**Proof** Firstly, according to Lemma 3.1, it can be known that $\Delta_{k,(j,m)} = tr(M_1 \times I_k \times \tau_k \times I_k \times M_2 \times M_2^{-1} \times Q_j \times \tau_{j,m} \times Q_j \times M_1^{-1}) = tr(I_k \times \tau_k \times I_k \times Q_j \times \tau_{j,m} \times Q_j)$. Then according to Lemma 3.2, since the diagonal elements of $I_k$, $\tau_k$, $Q_j$, $\tau_{j,m}$ are $(1, 1, 1, \cdots, 1, 0)$, $(r_k a_0, r_k a_1, r_k a_2, \ldots, r_k a_n, 0)$, $(1, 1, 1, \cdots, 1, 0)$, $(r_{j,m}, r_{j,m} T_{j,m}, r_{j,m} T_{j,m}^2, \ldots, r_{j,m} T_{j,m}^n, 0)$, respectively, it can be known that $tr(I_k \times \tau_k \times I_k \times Q_j \times \tau_{j,m} \times Q_j) = (1, 1, 1, \cdots, 1, 0) \circ (r_k a_0, r_k a_1, r_k a_2, \ldots, r_k a_n, 0) \circ (1, 1, 1, \cdots, 1, 0) \circ (r_{j,m}, r_{j,m} T_{j,m}, r_{j,m} T_{j,m}^2, \ldots, r_{j,m} T_{j,m}^n, 0) = r_k r_{j,m} f_{task_k}(T_{j,m})$. Since $r_k$ and $r_{j,m}$ do not equal to 0, if $\Delta_{k,(j,m)} = 0$, then $f_{task_k}(T_{j,m}) = 0$. Thus, Lemma 3.3 is proven.

### 3.3.3 Extension and Discussion

In order to make PPTA more suitable for real mobile crowdsensing task allocation scenarios, we extend PPTA as follows to support more functions, such as conjunctive task allocation, Top-z task allocation, task allocation with access control, and task recovery.

#### 3.3.3.1 Privacy-Preserving Conjunctive Task Allocation

In practical scenarios, data requesters may want to find users who have interests in multiple tasks at the same time. This way of task allocation is called conjunctive task allocation. For example, a doctor wants to find users who have both a cold and fever to collect their health data. PPTA can efficiently support privacy-preserving conjunctive task allocation. Specially, given task data $\{T_{j,m}\}_{m=1}^t$, $u_j$ chooses a sequences of random positive numbers $\{r_{j,m}\}_{m=1}^t$ and generates a lower triangular random matrix $\tau_j$ with main diagonal elements $(\sum_{m=1}^t r_{j,m}, \sum_{m=1}^t r_{j,m} T_{j,m}, \sum_{m=1}^t r_{j,m} T_{j,m}^2, \ldots, \sum_{m=1}^t r_{j,m} T_{j,m}^n, 0)$. Then, $u_j$

performs the trapdoor generation phase to obtain $T[\tau_j]$. The mobile crowdsensing platform performs the trapdoor transformation phase to obtain $RT[\tau_j]$. Next, the mobile crowdsensing platform performs the query phase as follows:

$$
\begin{aligned}
\Delta_{k,j} &= tr(RE[\tau_k] \times RT[\tau_j]) \\
&= tr(M_1 \times I_k \times \tau_k \times I_k \times M_2 \times M_2^{-1} \times Q_j \times \tau_j \times Q_j \times M_1^{-1}) \\
&= tr(M_1 \times I_k \times \tau_k \times I_k \times Q_j \times \tau_j \times Q_j \times M_1^{-1}) \\
&= r_k \sum_{m=1}^{t} f_{task_k}(T_{j,m}).
\end{aligned}
\tag{3.16}
$$

Since $r_k$ and $r_{j,m}$ are random positive number, it can be known that if and only if $\{f_{task_k}(T_{j,m}) = 0\}_{m=1}^{t}$, then $\Delta_{k,j} = 0$. That is, if and only if the sensing user $u_k$ satisfies all task requirements of $u_j$ simultaneously, $\Delta_{k,j} = 0$.

### 3.3.3.2 Privacy-Preserving Top-z Task Allocation

In Top-z task allocation scenarios, data requesters want to find users who are interested in $z$ or more tasks, where $z$ is a preset threshold. PPTA can efficiently support privacy-preserving Top-z task allocation. Specifically, the data requester generates $t > z$ trapdoors $\{RT[\tau_{j,m}]\}_{m=1}^{t}$ for its interested task $\{T_{j,m}\}_{m=1}^{t}$. The mobile crowdsensing platform selects the task ciphertext of user $u_k$ to calculate $\Delta_{k,(j,m)}$, and calculates $\Delta_{k,(j,0)}, \Delta_{k,(j,1)}, \dots, \Delta_{k,(j,t)}$ to obtain $u_k$'s task matching score $score_{k,j}$. If $score_{k,j} \geq z$, it means that the sensing user $u_k$ meets the task requirements of the data requester.

### 3.3.3.3 Privacy-Preserving Task Allocation with Access Control

In mobile crowdsensing task allocation scenarios, in order to further improve the accuracy of data collection, the mobile crowdsensing platform may wish to find sensing users with certain identities. For example, in order to know the best route for a destination area, data requesters are more inclined to find drivers active in the destination area to provide travel advice for them; in order to study the development trend of an epidemic on campus, medical institutions are more inclined to find a group of students with symptoms of the epidemic. PPTA can effectively support privacy-preserving task allocation with access control after a simple extension of its original design.

Specifically, in the initialization stage, the trusted center expands the dimension of the matrix from $(n + 2) \times (n + 2)$ to $(n + d + 3) \times (n + d + 3)$, where $d$ represents the number of user identities that the data requester wants to find. In the task encryption phase, the sensing user $u_k$ first obtains its identity $R_k$ from the identity dictionary *Role* and specifies the task information $\{T_{k,m}\}_{m=1}^{l}$ that $u_k$ is interested in. Then, $u_k$ chooses random positive numbers $r_{k,1}$ and $r_{k,2}$ and generates a $(n+d+3) \times (n+d+3)$-dimensional lower triangular random matrix $\tau_k$ with main diagonal elements $(r_{k,1}, r_{k,1}R_k, \ldots, r_{k,1}R_k^d, r_{k,2}a_0, r_{k,2}a_1, \ldots, r_{k,2}a_n, 0)$. In the task transformation phase, the mobile crowdsensing platform generates re-encryption ciphertext $RE[\tau_k]$. In the trapdoor generation phase, the data requester $u_j$ first specifies the identities $\{R_{j,m}\}_{m=1}^{t}$ that $u_j$ wants to find. Next, $u_j$ chooses meaningless random numbers $\{R_{j,m} \notin Role(m)\}_{m=t+1}^{d}$ and random number $b_d \neq 0$ and generates a polynomial function $f_{role_j}(x) = b_d(x - R_{j,1})(x - R_{j,2}) \cdots (x - Rj, d)$. Then, $u_j$ chooses random positive numbers $r_{j,1}$ and $r_{j,2}$ and generates a $(n+d+3) \times (n+d+3)$-dimensional lower triangular random matrix $\tau_{j,m}$ with main diagonal elements $(r_{j,1}b_0, r_{j,1}B_1, \ldots, r_{j,1}b_d, r_{j,2}, r_{j,2}T_{j,m}, \ldots, r_{j,2}T_{j,m}^n, 0)$. At last, $u_j$ encrypts $\tau_{j,m}$ as $T[\tau_{j,m}]$. In the trapdoor transformation phase, the mobile crowdsensing platform generates re-encryption trapdoor $RT[\tau_{j,m}]$. In the query phase, the mobile crowdsensing platform computes $\Delta_{k,(j,m)}$ as follows:

$$
\begin{aligned}
\Delta_{k,(j,m)} &= tr(RE[\tau_k] \times RT[\tau_{j,m}]) \\
&= r_{k,1}r_{j,1}f_{role_j}(R_k) + r_{k,2}r_{j,2}f_{task_k}(T_{j,m}).
\end{aligned}
\tag{3.17}
$$

Since $r_{k,1}, r_{k,2}, r_{j,1}$, and $r_{j,2}$ are random positive number, it can be known that if and only if $f_{role_j}(R_k) = 0$ and $f_{task_k}(T_{j,m}) = 0$ simultaneously, then $\Delta_{k,(j,m)} = 0$. That is, if and only if the sensing user satisfies the identity and task requirements that $u_j$ requires, then $\Delta_{k,(j,m)} = 0$.

### 3.3.3.4  Privacy-Preserving Task Recovery

In PPTA, data requesters can find sensing users that meet their task requirements. However, in practical applications, the data requester may not be able to communicate with the sensing users to inform them of the task information. After a simple extension, PPTA can efficiently support task recovery under privacy preservation. That is, PPTA allows the sensing users to know the task information required by the data requester but does not allow the mobile crowdsensing platform to obtain the specific task information.

Specifically, in the trapdoor phase, the data requester chooses a sequence of random numbers $\{r'_{j,i}\}_{i=0}^{n}$ and generates a $(n+2) \times (n+2)$-dimensional lower triangular random matrix $\tau'_{j,m}$ with main diagonal elements $(r'_{j,0}T_{j,m}, r'_{j,1}, \ldots, r'_{j,n}, 0)$. Next, $u_j$ encrypts $\tau'_{j,m}$ as $T[\tau'_{j,m}]$. In the trapdoor transformation phase, the mobile crowdsensing platform generates $RT[\tau'_{j,m}]$. In the query phase, if $\Delta_{k,(j,m)} = 0$, the mobile crowdsensing platform computes $\Delta'_{k,(j,m)}$ as follows:

$$\Delta'_{k,(j,m)} = tr(RE[\tau_k] \times RT[\tau'_{j,m}])$$

$$= r_k a_0 r'_{j,0} T_{j,m} + \sum_{i=1}^{n} r_k r'_{j,i} a_i. \tag{3.18}$$

Next, the mobile crowdsensing platform returns $\Delta'_{k,(j,m)}$ and $\{r'_{j,i}\}_{i=0}^{n}$ to sensing user $u_k$. Then, $u_k$ determines the specific value of $T_{j,m}$ that the data requester requires by checking whether there is a value in $\{T_{k,m}\}_{m=1}^{l}$, which satisfies

$$\frac{\Delta'_{k,(j,m)} - \sum_{i=1}^{n} r_k r'_{j,i} a_i}{r'_{j,0}} == r_k a_0 T_{k,m}. \tag{3.19}$$

Since each sensing user has different $r_k$ and $\{a_i\}_{i=0}^{n}$, and each data requester has different $\{r'_{j,i}\}_{i=0}^{n}$, the result $\Delta'_{k,(j,m)}$ obtained by the mobile crowdsensing platform is completely random to the mobile crowdsensing platform. That is, the mobile crowdsensing platform cannot compute or infer the sensitive information about task information of sensing users and query information of data requesters.

## 3.4  Security Analysis

This section analyzes the privacy preservation capability of PPTA under the given security model. Since PPTA utilizes a similar encryption method in the task encryption and trapdoor generation phases and also adopts a similar re-encryption method in the task transformation and trapdoor transformation phases, this section will take task encryption and task transformation as examples to analyze the security of PPTA.

### 3.4.1  Security Under Passive Attack

In PPTA, the passive attack means that the adversary can only obtain the user's ciphertext information. We first give the game between the adversary $\mathcal{A}$ and the challenger $C$, as shown in Fig. 3.3, and then we give theorems and related proofs to demonstrate the security of PPTA.

Based on $Passive_{A,PPTA}(\lambda)$, we have the following definitions.

1.**Init:** Given a security parameter $\lambda$, the adversary $\mathcal{A}$ generates two databases $DB_0 = (D_{0,1}, D_{0,2}, \cdots, D_{0,t})$ and $DB_1 = (D_{1,1}, D_{1,2}, \cdots, D_{1,t})$, where $D_{k,m}$ ($k \in 0, 1, m \in 1, 2, 3, \cdots, t$) represents a $(n + 2)$-dimensional random vector and each element in $D_{k,m}$ ranges from 1 to $2^{\lambda}$. Then, $\mathcal{A}$ sends $DB_0$ and $DB_1$ to the challenger $C$.

2.**Setup:** The challenger $C$ performs the system initialization phase in PPTA to generate the master secret key, encryption key, and re-encryption key.

3.**Challenge:** According to the data received from **Init**, the challenger $C$ tosses a coin to choose $b \in 0, 1$, and extends $D_{b,m}$ ($m \in 1, 2, 3, \cdots, t$) in $DB_b$ to a $(n + 2) \times (n + 2)$-dimensional lower triangular random matrix $\tau_{b,m}$ with main diagonal elements $D_{b,m}$. Next, the challenger $C$ performs the task encryption phase to obtain $E[\tau_{b,m}]$ and performs the task transformation phase to obtain $RE[\tau_{b,m}]$. Then, $C$ returns $E[\tau_{b,m}]$ and $RE[\tau_{b,m}]$ to the adversary $\mathcal{A}$.

4.**Guess:** The adversary $\mathcal{A}$ guesses that $b'$ is either 0 or 1.

5.**Output:** If $b' = b$, output 1. Otherwise, output 0.

**Fig. 3.3** Passive attack $Passive_{\mathcal{A},PPTA}$

**Definition 3.1** For any adversary $\mathcal{A}$ with probabilistic polynomial time (P.P.T), it has a negligible advantage in guessing $b' = b$ as follows:

$$|Pr(Passive_{A,PPTA}(\lambda) = 1) - \frac{1}{2}| \leq negl(\lambda), \qquad (3.20)$$

and then PPTA can resist the passive attack.

Based on Definition 3.1, we have the following theorem.

**Theorem 3.4** *PPTA can resist the passive attack.*

*Proof* Firstly, according to Fig. 3.3, it can be known that the adversary $\mathcal{A}$ knows the plaintext information and the ciphertext information returned from the challenger $C$. However, the adversary $\mathcal{A}$ does not know their correspondence. Since $\mathcal{A}$ does not know the specific data in $\tau_{b,m}$, $\mathcal{A}$ can only set $\tau_{b,m}$ as a $(n+2) \times (n+2)$-dimensional lower triangular random matrix. Then, based on the received ciphertext $E[\tau_{b,m}] = A_{k,1} \times \tau_{b,m} \times A_{k,2}$, $\mathcal{A}$ can conduct $(n+2) \times (n+2)$ equations. However, since there are $\frac{(n+1) \times (n+4)}{2} + 2(n+2) \times (n+2)$ unknowns, $\mathcal{A}$ cannot obtain any information from the equations. Similarly, $\mathcal{A}$ cannot recover $\tau_{b,m}$ from the re-encrypted task information $RE[\tau_{b,m}] = M_1 \times I_k \times \tau_{b,m} \times I_k \times M_2$. Thus, the ciphertext obtained by $\mathcal{A}$ is completely random for $\mathcal{A}$.

In summary, the advantage for $\mathcal{A}$ to guess $b' = b$ is $|Pr(Passive_{A,PPTA}(\lambda) = 1) - \frac{1}{2}| \leq negl(\lambda)$. Thus, Theorem 3.4 is proven.

### 3.4.2  Security Under Active Attack

In addition to the ciphertexts, an adversary with active attack capability may also obtain the plaintexts corresponding to the ciphertexts. In this section, we consider two active attack types: (1) known-plaintext attacks launched by the mobile crowdsensing platform and (2) chosen-plaintext attacks launched by the adversary (non-mobile crowdsensing platform). For the known-plaintext attack, the mobile crowdsensing platform may obtain some task plaintexts and the corresponding ciphertexts and launch attacks [46, 47] based on these plaintext-ciphertext pairs to obtain some users' keys or task information. Since PPTA utilizes a task dictionary (such as a secure hash function) to map task information to any random number, the mobile crowdsensing platform cannot determine the random number information corresponding to a task. Thus, the mobile crowdsensing platform cannot obtain the key information or the user's task privacy through the task matching equations. The related proofs are similar to the proofs in scheme [23]. To avoid repetition, this chapter omits the security proofs for the known-plaintext attack launched by the mobile crowdsensing platform. Then, we consider the chosen-plaintext attack launched by an adversary (a non-mobile crowdsensing platform). In this case, the adversary may choose some special plaintexts and observe their corresponding ciphertexts. For this kind of attack, we first simulate the game between the adversary $\mathcal{A}$ and the challenger $C$, as shown in Fig. 3.4. Then, we give the definition of PPTA under the active attack $Active_{\mathcal{A},PPTA}$. Finally, we give the theorem and its corresponding proof.

---

1.**Init:** Given a security parameter $\lambda$, the adversary $\mathcal{A}$ generates two databases $DB_0 = (D_{0,1}, D_{0,2}, \cdots, D_{0,t})$ and $DB_1 = (D_{1,1}, D_{1,2}, \cdots, D_{1,t})$, where $D_{k,m}(k \in 0,1, m \in 1,2,3,\cdots,t)$ represents a $(n+2)$-dimensional random vector and each element in $D_{k,m}$ ranges from 1 to $2^\lambda$.

2.**Setup:** The challenger $C$ performs the system initialization phase in PPTA to generate the master secret key, encryption key, and re-encryption key.

3.**Phase 1:** The adversary $\mathcal{A}$ sends $DB_l = (D_{l,1}, D_{l,2}, \cdots, D_{l,t})$ to the challenger $C$ in the $l$-th query. Next, $C$ extends the vector $DB_{l,m} (m \in 1,2,3,\cdots,t)$ in $DB_l$ to a $(n+2)\times(n+2)$-dimensional lower triangular random matrix $\tau_{l,m}$ with main diagonal elements $DB_{l,m}$. Then, $C$ performs the task encryption phase to obtain $E[\tau_{l,m}]$ and performs the task transformation phase to obtain $RE[\tau_{l,m}]$. Finally, $C$ returns $E[\tau_{l,m}]$ and $RE[\tau_{l,m}]$ to $\mathcal{A}$.

4.**Challenge:** The challenger $C$ tosses a coin to choose $b \in 0,1$, and extends $D_{b,m} (m \in 1,2,3,\cdots,t)$ in $DB_b$ to a $(n+2)\times(n+2)$-dimensional lower triangular random matrix $\tau_{b,m}$ with main diagonal elements $D_{b,m}$. Next, the challenger $C$ performs the task encryption phase to obtain $E[\tau_{b,m}]$ and performs the task transformation phase to obtain $RE[\tau_{b,m}]$. Then, $C$ returns $E[\tau_{b,m}]$ and $RE[\tau_{b,m}]$ to the adversary $\mathcal{A}$.

5.**Phase 2:** The adversary $\mathcal{A}$ repeats the **Phase 1**.

6.**Guess:** The adversary $\mathcal{A}$ guesses that $b'$ is either 0 or 1.

7.**Output:** If $b' = b$, output 1. Otherwise, output 0.

---

**Fig. 3.4** Active attack $Active_{A,PPTA}$

Based on $Passive_{A,PPTA}(\lambda)$, we have the following definitions.

**Definition 3.2** For any adversary $\mathcal{A}$ with probabilistic polynomial time (P.P.T), it has a negligible advantage in guessing $b' = b$ as follows:

$$|Pr(Active_{A,PPTA}(\lambda) = 1) - \frac{1}{2}| \leq negl(\lambda), \qquad (3.21)$$

and then PPTA can resist the active attack.

Based on Definition 3.2, we have the following theorem.

**Theorem 3.5** *PPTA can resist the active attack.*

***Proof*** In order to prove that PPTA can resist the active attack, we need to prove that it can resist the chosen plaintext attack. That is, we need to prove that the adversary is indistinguishable from $E[T_{0,m}]$, $E[T_{1,m}]$ (or $RE[T_{0,m}]$, $RE[T_{1,m}]$).

We take the data $D_{0,1}$ as an example to prove the theorem. Firstly, assuming that $D_{0,1} = (d_1, d_2, d_3, \ldots, d_{n+1}, 0)$ represents a random vector generated by the adversary $\mathcal{A}$. From the construction details of PPTA, $D_{0,1}$ is extended to a $(n+2) \times (n+2)$-dimensional lower triangular random matrix $\tau_{0,1}$ with main diagonal elements $(r_k d_1, r_k d_2, r_k d_3, \ldots, r_k d_{n+1}, 0)$ and then encrypted as $E[\tau_{0,1}] = A_{k,1} \times \tau_{0,1} \times A_{k,2}$, where $r_k \neq 0$. According to the principle of matrix multiplication, we assume that the product of $A_{k,1}$ and $\tau_{0,1}$ is $AT$, and the element $at_{i,j}(i, j \in 1, 2, \ldots, n+2)$ can be represented as

$$at_{i,j} = a_{i,1}^{(1)} t_{1,j} + a_{i,2}^{(1)} t_{2,j} + \cdots + a_{i,n+2}^{(1)} t_{n+2,j}$$
$$= \sum_{k=1}^{n+2} a_{i,k}^{(1)} t_{k,j}, \qquad (3.22)$$

where $a_{i,j}^{(1)}$ and $t_{i,j}$ are the elements of the $i$-th row and the $j$-th column in $A_{k,1}$ and $\tau_{0,1}$, respectively. Specifically, $t_{i,j}$ satisfies

$$\begin{cases} t_{i,j} = *, 1 \leq j < i \leq n+2, \\ t_{i,j} = r_k d_i, 1 \leq j = i \leq n+2, \\ t_{i,j} = 0, \text{other.} \end{cases}$$

For simplicity, we utilize $*$ to represent the random in $\tau_{0,1}$. Then, we utilize $\bar{at}_{i,j}(i, j \in 1, 2, 3, \ldots, n+2)$ to represent the element in $E[\tau_{0,1}]$ and compute them as

$$\bar{a}t_{i,j} = at_{i,1}a^{(2)}_{1,j} + at_{i,2}a^{(2)}_{2,j} + \cdots + at_{i,n+2}a^{(2)}_{n+2,j}$$

$$= \sum_{k=1}^{n+2} at_{i,k}a^{(1)}_{k,j} \tag{3.23}$$

$$= \sum_{k=1}^{n+2}(\sum_{z=1}^{n+2} a^{(1)}_{i,z}t_{z,k})a^{(2)}_{k,j},$$

where $\{a^{(2)}_{i,j}\}^{n+2}_{i=1,j=1}$ is the element in $A_{k,2}$. Since $a^{(1)}_{i,j}$ and $a^{(2)}_{i,j}$ are fixed number and $t_{i,j}(1 \leq j < i \leq n+2)$ is a random generated by the challenger $C$, it can be known that $\bar{a}t_{i,j}$ is a random related to $\{t_{i,j}\}_{1\leq j<i\leq n+2}$. It means that even if $\mathcal{A}$ can choose a different $D_{k,j}$ to launch a query in $C$, the ciphertexts $\mathcal{A}$ obtains each time are completely random to $\mathcal{A}$. Similarly, we can prove that $RE[\tau_{b,j}]$ is completely random to $\mathcal{A}$.

In summary, the advantage for $\mathcal{A}$ to guess $b' = b$ is $|Pr(Active_{A,PPTA}(\lambda) = 1) - \frac{1}{2}| \leq negl(\lambda)$. Thus, Theorem 3.5 is proven.

## 3.5  Performance Evaluation and Analysis

In this section, we theoretically and experimentally evaluate the performance of PPTA. For comparison, we consider existing privacy-preserving task allocation schemes EPTR [23] and PPTR [26]. In particular, we make some modifications to EPTR and PPTR according to the PPTA construction rules. Specifically, in EPTR, each task interest of the user is encrypted separately. This task encryption method will cause large storage overhead to the mobile crowdsensing platform. In addition, in their scheme, the mobile crowdsensing platform can infer some private information based on the query results. For example, if the query of the two data requesters finds the same task interest of the same sensing user, then the mobile crowdsensing platform knows that the two data requesters have the same task requirements. Therefore, to fairly compare the performance of the schemes, we use the coefficients of a polynomial function to generate task ciphertexts for sensing users in EPTR. For PPTR, we use each element of the matrix to represent a task and randomly generate a matrix to perturb the task matrix. For simplicity, we assume that each sensing user has only one task ciphertext. Table 3.2 shows the symbols used in the theoretical analysis and their corresponding meanings.

**Table 3.2** Notations and their corresponding meanings used in the theoretical analysis of PPTA

| Notation | Meaning |
|---|---|
| $W$ | Number of task types in the task dictionary |
| $K$ | Number of sensing users |
| $\lambda$ | Size of an element of the matrix and vector |
| $T_A$ | Time of the matrix multiplication operation in PPTA |
| $S_A$ | Size of a matrix in PPTA |
| $T_E$ | Time of the vector multiplication in EPTR |
| $S_E$ | Size of a vector in EPTR |
| $T_R$ | Time of the matrix multiplication in PPTR |
| $S_R$ | Size of a matrix in PPTR |

## 3.5.1  Theoretical Analysis

In this section, we analyze the computational cost and communication overhead in PPTA. In PPTA, a sensing user first needs to perform two matrix multiplication operations in the task encryption phase to encrypt the task information that he/she is interested in. Then, the mobile crowdsensing platform needs to perform two matrix multiplication operations to re-encrypt the task ciphertext. In the trapdoor generation phase, a data requester needs to perform two matrix multiplication operations to encrypt the task information. Similarly, the mobile crowdsensing platform needs to perform two matrix multiplication operations in the trapdoor re-encryption phase to re-encrypt a task ciphertext. In the query phase, the mobile crowdsensing platform needs to perform $K$ matrix multiplication operations, where $K$ represents the number of sensing users. The theoretical analysis results in terms of computational cost and communication overhead are summarized in Table 3.3. Since $T_A = O((n+1)^3)$, $T_E = O((n+1)^2)$, $T_R = O((|\sqrt{W}|)^3)$, and $W \gg n$, it can be seen that $T_E < T_A < T_R$. Since $S_A = \lambda(n+1)^2$, $S_E = \lambda(n+1)$, $S_R = \lambda(|\sqrt{W}|)^2$, and $W \gg n$, it can be seen that $S_E < S_A < S_R$.

**Table 3.3** Theoretical analysis of computational cost and communication overhead

| Overhead | Phase | Entity | EPTR | PPTR | PPTA |
|---|---|---|---|---|---|
| Computational cost | Task encryption | Sensing user | $2T_E$ | $T_R$ | $2T_A$ |
| | Task transformation | Mobile crowdsensing platform | $2T_E$ | – | $2T_A$ |
| | Trapdoor generation | Data requester | $2T_E$ | $T_R$ | $2T_A$ |
| | Trapdoor transformation | Mobile crowdsensing platform | $2T_E$ | – | $2T_A$ |
| | Query | Mobile crowdsensing platform | $2KT_E$ | $KT_R$ | $KT_A$ |
| Communication overhead | Task encryption | Sensing user | $2S_E$ | $S_R$ | $S_A$ |
| | Task transformation | Mobile crowdsensing platform | – | – | – |
| | Trapdoor generation | Data requester | $2S_E$ | $S_R$ | $S_A$ |
| | Trapdoor transformation | Mobile crowdsensing platform | – | – | – |
| | Query | Mobile crowdsensing platform | – | – | – |

## *3.5.2  Performance Evaluation*

**Experimental Environment and Implementation**  Since the lack of large-scale datasets is related to mobile crowdsensing tasks, we implement a privacy-preserving task allocation system based on real mobile crowdsensing scenarios. Specifically, we use a laptop with 16.0 GB of RAM and an Intel i5 processor as the mobile crowdsensing platform and an Android phone with 4.0 GB of RAM as the sensing user and data requester. The number of tasks in the task dictionary is set from 1000 to 10,000, where each task is mapped to a random number in $[1, 2^{128}]$. In order to generate a large-scale task database, we let sensing users randomly select tasks and send them to the mobile crowdsensing platform after encryption. To perform the query, we let the data requester randomly select a task to generate a trapdoor and send it to the mobile crowdsensing platform. We implement schemes PPTA, EPTR [23], and PPTR [26] on Android phones and laptops. Among them, we implement task encryption and trapdoor generation steps on Android phones in Java and use Python to implement task re-encryption, trapdoor re-encryption, and task query steps on laptops.

**Experimental Result**  Figure 3.5 presents the computational cost of the sensing user in the task encryption phase, where the number of tasks required by the data requester ranges from 5 to 20. It can be seen that as the number of tasks increases, the computational cost of both PPTA and EPTR increases. This is because the number of tasks increases, and the matrix dimension increases, which leads to an increase in the computation time of matrix multiplication. Compared with EPTR, PPTA takes more time to perform task encryption. Specifically, when the number of tasks is 20, PPTA takes 0.177 ms, while EPTR takes only 0.076 ms. This is because the task encryption in PPTA is matrix-matrix multiplication, while in EPTR, it is matrix-vector multiplication. Similarly, in Fig. 3.6, PPTA requires more time in the trapdoor generation phase on the data requester side than EPTR. Although EPTR is more efficient than PPTA, it should be pointed out that EPTR achieves such high computational efficiency by sacrificing certain security. When the number of required tasks is 20, the computational cost of PPTA in both task encryption and trapdoor generation phases is less than 0.2 ms, which proves the high efficiency of PPTA on the user side.

Figures 3.7 and 3.8 show the computational cost of PPTR in the task encryption phase and the trapdoor generation phase, respectively, where the number of tasks in the task dictionary ranges from 2000 to 10,000. It can be seen that with the increase in task types, the computational cost of PPTR also increases gradually. Specifically, when the task type is 2000, PPTR needs about 0.23 milliseconds to perform the data encryption operation on the sensing user side and the data requester side, respectively. When the number of tasks is 10,000, PPTR takes about 1.59 milliseconds on the sensing user side and the data requester side, respectively. Due to the use of the polynomial function, PPTA cannot be affected by the change in the

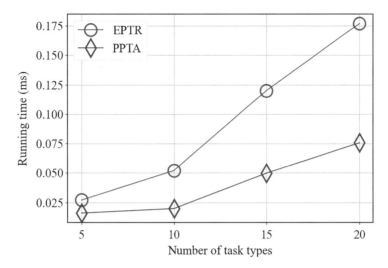

**Fig. 3.5** Computational cost of the sensing user in the task encryption phase

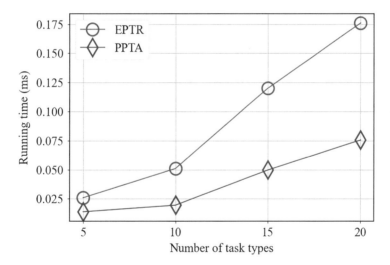

**Fig. 3.6** Computational cost of the data requester in the trapdoor generation phase

number of tasks in the task dictionary. Thus, PPTA saves more computational costs in the data encryption and trapdoor generation phases compared to PPTR.

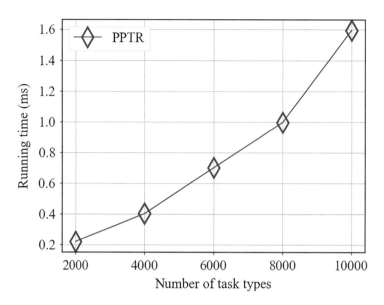

**Fig. 3.7** Computational cost in the task encryption phase

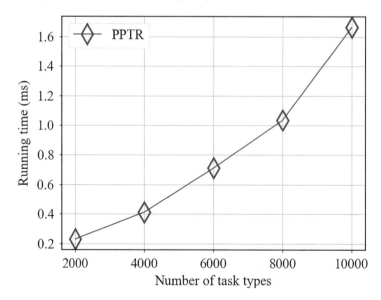

**Fig. 3.8** Computational cost in the trapdoor generation phase

Figures 3.9 and 3.10 show the computational cost of PPTA and EPTR in the task re-encryption and trapdoor re-encryption phases, respectively, where the number of tasks required is set to 20. That is, the matrix dimension is set to 21. It can be seen that PPTA can efficiently perform task re-encryption and trapdoor re-encryption

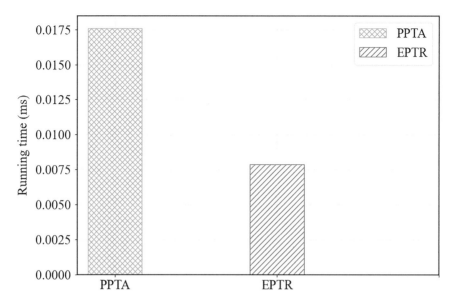

**Fig. 3.9**  Computational cost in the task re-encryption phase

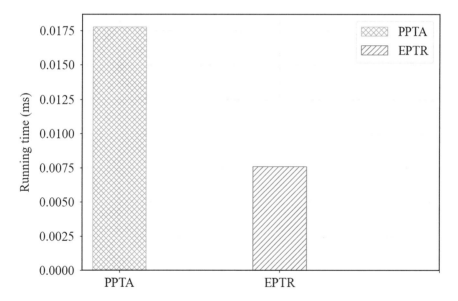

**Fig. 3.10**  Computational cost in the trapdoor re-encryption phase

operations. Specifically, PPTA only needs about 0.018 milliseconds to re-encrypt a task ciphertext or a trapdoor ciphertext.

Figure 3.11 shows the computational cost of PPTA, EPTR, and PPTR in the query phase, where the number of sensing users ranges from 10,000 to 100,000.

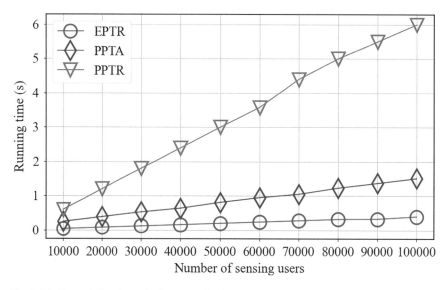

**Fig. 3.11** Computational cost in the query phase

Among them, for PPTR, the number of tasks in the task dictionary is set to 2000. It can be seen that as the number of sensing users increases, the computational cost of all schemes also increases. Compared with the other two schemes, PPTA can achieve better computational efficiency. For example, when the number of sensing users is 100,000, PPTR takes 6.03 seconds to complete all query operations, while PPTA takes only 1.56 seconds. The experimental results demonstrate the efficiency of PPTA in the query phase.

## 3.6  Summary

Content-based task allocation can help data requesters find sensing users who meet their task requirements, but task information that users are interested in may leak users' privacy. In this chapter, we study the security and privacy issues of privacy-preserving content-based task allocation in mobile crowdsensing scenarios and propose a privacy-preserving content-based task allocation scheme. Specifically, we first design a basic privacy-preserving task allocation scheme (PPTA) based on random matrix multiplication and polynomial function. This scheme enables the mobile crowdsensing platform to find the sensing users who meet the task requirements of the data requester without compromising the user's task and query privacy. Then, we further expand PPTA to support more functions, such as conjunctive task allocation, Top-z task allocation, task allocation with access control, and task recovery, which ensures the versatility and practicability of the scheme. Security analysis proves that PPTA can effectively preserve users' task privacy and query privacy. On the basis of the scheme in this chapter, it is a meaningful research direction to study a privacy-preserving content-based allocation scheme with nonlinear query efficiency.

# References

1. Yan, X., Ma, M.: A privacy-preserving handover authentication protocol for a group of MTC devices in 5g networks. Comput. Secur. **116**, 102601 (2022)
2. Satheeshkumar, R., Saini, K., Daniel, A., Khari, M.: Chapter seventeen-5G-communication in healthcare applications. Adv. Comput. **127**, 485–506 (2022)
3. Shehab, M.J., Kassem, I., Kutty, A.A., Kucukvar, M., Onat, N., Khattab, T.: 5G networks towards smart and sustainable cities: a review of recent developments, applications and future perspectives. IEEE Access **10**, 2987–3006 (2022)
4. Letaief, K.B., Shi, Y., Lu, J., Lu, J.: Edge artificial intelligence for 6G: vision, enabling technologies, and applications. IEEE J. Sel. Areas Commun. **40**(1), 5–36 (2022)
5. Du, X., Zhou, Z., Zhang, Y., Rahman, T.: Energy-efficient sensory data gathering based on compressed sensing in IoT networks. J. Cloud Comput. **9**, 19 (2020)
6. Liu, L., Chen, B., Ma, H.: SDCN: sensory data-centric networking for building the sensing layer of IoT. ACM Trans. Sens. Netw. **17**(1), 6:1–6:25 (2020)
7. Yurur, O., Liu, C.H., Moreno, W.: Modeling battery behavior on sensory operations for context-aware smartphone sensing. Sensors **15**(6), 12323–12341 (2015)
8. Tong, Y., Zhou, Z., Zeng, Y., Chen, L., Shahabi, C.: Spatial crowdsourcing: a survey. VLDB J. **29**(1), 217–250 (2020)
9. Kazai, G., Kamps, J., Koolen, M., Milic-Frayling, N.: Crowdsourcing for book search evaluation: impact of hit design on comparative system ranking. In: Ma, W.-Y., Nie, J.-Y., Baeza-Yates, R., Chua, T.-S., Croft, W.B. (eds.) Proceeding of the 34th International ACM SIGIR Conference on Research and Development in Information Retrieval, SIGIR 2011, Beijing, July 25–29, 2011, pp. 205–214. ACM (2011)
10. Blanco, R., Halpin, H., Herzig, D.M., Mika, P., Pound, J., Thompson, H.S., Tran Duc, T.: Repeatable and reliable search system evaluation using crowdsourcing. In: Ma, W.-Y., Nie, J.-Y., Baeza-Yates, R., Chua, T.-S., Croft, W.B. (eds.) Proceeding of the 34th International ACM SIGIR Conference on Research and Development in Information Retrieval, SIGIR 2011, Beijing, July 25–29, 2011, pp. 923–932. ACM (2011)
11. Zhang, C., Zhu, L., Xu, C., Liu, X., Sharif, K.: Reliable and privacy-preserving truth discovery for mobile crowdsensing systems. IEEE Trans. Dependable Secur. Comput. **18**(3), 1245–1260 (2021)
12. Stowell, E., O'Leary, T.K., Kimani, E., Paasche-Orlow, M.K., Bickmore, T., Parker, A.G.: Investigating opportunities for crowdsourcing in church-based health interventions: A participatory design study. In: Bernhaupt, R., 'Floyd' Mueller, F., Verweij, D., Andres, J., McGrenere, J., Cockburn, A., Avellino, I., Goguey, A., Bjøn, P., Zhao, S., Samson, B.P., Kocielnik, R. (eds.) CHI '20: CHI Conference on Human Factors in Computing Systems, Honolulu, HI, April 25–30, 2020, pp. 1–12. ACM (2020)
13. Wu, T., Chen, L., Hui, P., Zhang, C.J., Li, W.: Hear the whole story: towards the diversity of opinion in crowdsourcing markets. Proc. VLDB Endow. **8**(5), 485–496 (2015)
14. Borromeo, R.M., Laurent, T., Toyama, M., Alsayasneh, M., Amer-Yahia, S., Leroy, V.: Deployment strategies for crowdsourcing text creation. Inf. Syst. **71**, 103–110 (2017)
15. Ji, S., Shao, J., Agun, D., Yang, T.: Privacy-aware ranking with tree ensembles on the cloud. In: Collins-Thompson, K., Mei, Q., Davison, B.D., Liu, Y., Yilmaz, E. (eds.) The 41st International ACM SIGIR Conference on Research & Development in Information Retrieval, SIGIR 2018, Ann Arbor, MI, July 08–12, 2018, pp. 315–324. ACM (2018)

16. Wu, H., Wang, L., Xue, G., Tang, J., Yang, D.: Enabling data trustworthiness and user privacy in mobile crowdsensing. IEEE/ACM Trans. Netw. **27**(6), 2294–2307 (2019)
17. Wang, B., Li, M., Xiong, L.: Fastgeo: Efficient geometric range queries on encrypted spatial data. IEEE Trans. Dependable Secur. Comput. **16**(2), 245–258 (2019)
18. Lai, S., Patranabis, S., Sakzad, A., Liu, J.K., Mukhopadhyay, D., Steinfeld, R., Sun, S., Liu, D., Zuo, C.: Result pattern hiding searchable encryption for conjunctive queries. In: Lie, D., Mannan, M., Backes, M., Wang, X. (eds.) Proceedings of the 2018 ACM SIGSAC Conference on Computer and Communications Security, CCS 2018, Toronto, ON, October 15–19, 2018, pp. 745–762. ACM (2018)
19. Jarecki, S., Jutla, C., Krawczyk, H., Rosu, M.-C., Steiner, M.: Outsourced symmetric private information retrieval. In: Sadeghi, A.-R., Gligor, V.D., Yung, M. (eds.) 2013 ACM SIGSAC Conference on Computer and Communications Security, CCS'13, Berlin, November 4–8, 2013, pp. 875–888. ACM (2013)
20. Tang, Q.: Nothing is for free: Security in searching shared and encrypted data. IEEE Trans. Inf. Forensics Secur. **9**(11), 1943–1952 (2014)
21. Zhang, W., Lin, Y., Xiao, S., Wu, J., Zhou, S.: Privacy preserving ranked multi-keyword search for multiple data owners in cloud computing. IEEE Trans. Comput. **65**(5), 1566–1577 (2016)
22. Tang, W., Zhang, K., Ren, J., Zhang, Y., (Sherman) Shen, X.: Privacy-preserving task recommendation with win-win incentives for mobile crowdsourcing. Inf. Sci. **527**, 477–492 (2020)
23. Shu, J., Jia, X., Yang, K., Wang, H.: Privacy-preserving task recommendation services for crowdsourcing. IEEE Trans. Serv. Comput. **14**(1), 235–247 (2021)
24. Wong, W.K., Cheung, D.W.L., Kao, B., Mamoulis, N.: Secure kNN computation on encrypted databases. In: Çetintemel, U., Zdonik, S.B., Kossmann, D., Tatbul, N. (eds.) Proceedings of the ACM SIGMOD International Conference on Management of Data, SIGMOD 2009, Providence, USA, June 29–July 2, 2009, pp. 139–152. ACM (2009)
25. Gu, C., Gu, J.: Known-plaintext attack on secure knn computation on encrypted databases. Secur. Commun. Netw. **7**(12), 2432–2441 (2014)
26. Ni, J., Zhang, K., Xia, Q., Lin, X., Shen, X.S.: Enabling strong privacy preservation and accurate task allocation for mobile crowdsensing. IEEE Trans. Mob. Comput. **19**(6), 1317–1331 (2020)
27. Yuen, M.-C., King, I., Leung, K.-S.: Taskrec: a task recommendation framework in crowdsourcing systems. Neural Process. Lett. **41**(2), 223–238 (2015)
28. Yang, S., Han, K., Zheng, Z., Tang, S., Wu, F.: Towards personalized task matching in mobile crowdsensing via fine-grained user profiling. In: 2018 IEEE Conference on Computer Communications, INFOCOM 2018, Honolulu, HI, April 16–19, 2018, pp. 2411–2419. IEEE (2018)
29. Karaliopoulos, M., Koutsopoulos, I., Titsias, M.K.: First learn then earn: optimizing mobile crowdsensing campaigns through data-driven user profiling. In: Dressler, F., auf der Heide, F.M. (eds.) Proceedings of the 17th ACM International Symposium on Mobile Ad Hoc Networking and Computing, MobiHoc 2016, Paderborn, July 4–8, 2016, pp. 271–280. ACM (2016)
30. Kobren, A., Tan, C.H., Ipeirotis, P., Gabrilovich, E.: Getting more for less: Optimized crowdsourcing with dynamic tasks and goals. In: Gangemi, A., Leonardi, S., Panconesi, A. (eds.) Proceedings of the 24th International Conference on World Wide Web, WWW 2015, Florence, May 18–22, 2015, pp. 592–602. ACM (2015)
31. Wang, Z., Tan, R., Hu, J., Zhao, J., Wang, Q., Xia, F., Niu, X.; Heterogeneous incentive mechanism for time-sensitive and location-dependent crowdsensing networks with random arrivals. Comput. Netw. **131**, 96–109 (2018)
32. Tong, Y., She, J., Ding, B., Wang, L., Chen, L.: Online mobile micro-task allocation in spatial crowdsourcing. In: t32nd IEEE International Conference on Data Engineering, ICDE 2016, Helsinki, May 16–20, 2016, pp. 49–60. IEEE Computer Society (2016)

33. Gong, Y., Guo, Y., Fang, Y.: A privacy-preserving task recommendation framework for mobile crowdsourcing. In: IEEE Global Communications Conference, GLOBECOM 2014, Austin, TX, December 8–12, 2014, pp. 588–593. IEEE (2014)
34. Shu, J., Jia, X.: Secure task recommendation in crowdsourcing. In: 2016 IEEE Global Communications Conference, GLOBECOM 2016, Washington, DC, December 4–8, 2016, pp. 1–6. IEEE (2016)
35. Shu, J., Liu, X., Jia, X., Yang, K., Deng, R.H.: Anonymous privacy-preserving task matching in crowdsourcing. IEEE Int. Things J. 5(4), 3068–3078 (2018)
36. Ni, J., Zhang, K., Lin, X., Xia, Q., Shen, X.S.: Privacy-preserving mobile crowdsensing for located-based applications. In: IEEE International Conference on Communications, ICC 2017, Paris, May 21–25, 2017, pp. 1–6. IEEE (2017)
37. Xu, J., Cui, B., Shi, R., Feng, Q.: Outsourced privacy-aware task allocation with flexible expressions in crowdsourcing. Future Gener. Comput. Syst. 112, 383–393 (2020)
38. Yin, H., Xiong, Y., Deng, T., Deng, H., Zhu, P.: A privacy-preserving and identity-based personalized recommendation scheme for encrypted tasks in crowdsourcing. IEEE Access 7, 138857–138871 (2019)
39. Hao, J., Huang, C., Chen, G., Xian, M., Shen, X.S.: Privacy-preserving interest-ability based task allocation in crowdsourcing. In: 2019 IEEE International Conference on Communications, ICC 2019, Shanghai, May 20–24, 2019, pp. 1–6. IEEE (2019)
40. Wang, L., Yang, D., Han, X., Wang, T., Zhang, D., Ma, X.: Location privacy-preserving task allocation for mobile crowdsensing with differential geo-obfuscation. In: Barrett, R., Cummings, R., Agichtein, E., Gabrilovich, E. (eds.) Proceedings of the 26th International Conference on World Wide Web, WWW 2017, Perth, April 3–7, 2017, pp. 627–636. ACM (2017)
41. Alamer, A., Ni, J., Lin, X., Shen, X.: Location privacy-aware task recommendation for spatial crowdsourcing. In: 9th International Conference on Wireless Communications and Signal Processing, WCSP 2017, Nanjing, October 11–13, 2017, pp. 1–6. IEEE (2017)
42. Zhang, J., Zhang, Q., Ji, S.: A fog-assisted privacy-preserving task allocation in crowdsourcing. IEEE Int. Things J. 7(9), 8331–8342 (2020)
43. Hu, Y., Shen, H., Bai, G., Wang, T.: Privacy-preserving task allocation for edge computing enhanced mobile crowdsensing. In: Vaidya, J., Li, J. (eds.) Algorithms and Architectures for Parallel Processing—18th International Conference, ICA3PP 2018, Guangzhou, November 15–17, 2018, Proceedings, Part IV. Lecture Notes in Computer Science, vol. 11337, pp. 431–446. Springer (2018)
44. Liu, C., Zhu, L., Wang, M., Tan, Y.A.: Search pattern leakage in searchable encryption: attacks and new construction. Inf. Sci. 265, 176–188 (2014)
45. Cao, N., Wang, C., Li, M., Ren, K., Lou, W.: Privacy-preserving multi-keyword ranked search over encrypted cloud data. IEEE Trans. Parallel Distributed Syst. 25(1), 222–233 (2014)
46. Yao, B., Li, F., Xiao, X.: Secure nearest neighbor revisited. In: Jensen, C.S., Jermaine, C.M., Zhou, X. (eds.) 29th IEEE International Conference on Data Engineering, ICDE 2013, Brisbane, April 8–12, 2013, pp. 733–744. IEEE Computer Society (2013)
47. Lin, W., Wang, K., Zhang, Z., Chen, H.: Revisiting security risks of asymmetric scalar product preserving encryption and its variants. In: Lee, K., Liu, L. (eds.) 37th IEEE International Conference on Distributed Computing Systems, ICDCS 2017, Atlanta, GA, June 5–8, 2017, pp. 1116–1125. IEEE Computer Society (2017)

# Chapter 4
# Privacy-Preserving Location-Based Task Allocation

**Abstract** Chapter 3 discusses privacy-preserving task allocation based on task content. In mobile crowdsensing (MCS), since tasks are usually closely related to locations, data requesters may wish to find sensing users located within a certain geometric range to complete the task. Therefore, this chapter will study the privacy-preserving task allocation based on geometric range. Existing privacy-preserving geometric range query schemes either cannot achieve query in arbitrary geometric range or rely on trusted data centers to process location data. To address the challenges, this chapter proposes a privacy-preserving task allocation scheme GPTA that supports arbitrary geometric range queries. Specifically, this chapter firstly proposes GPTA-L with linear query efficiency based on the techniques of polynomial fitting and random matrix multiplication. In order to further improve query efficiency, this chapter studies the query history of data requesters and designs a nonlinear query efficiency scheme GPTA-F based on geometric properties. Security analysis proves that the proposed scheme can effectively preserve the user's location privacy and query privacy. Experiments based on the mobile crowdsensing applications verify that the proposed scheme has high computational and communication efficiency.

## 4.1 Introduction

This section describes the overview, related works, and preliminary of this chapter.

### 4.1.1 Overview

Mobile crowdsensing applications, such as traffic monitoring [1, 2], environment sensing [3, 4], noise detection [5, 6], etc., are usually closely related to location information. In order to improve the quality of data collection and also to speed up the efficiency of data collection, data requesters may wish to find sensing users located within a certain geometric range for completing data collection tasks. For example, Alice wants to stay at the Friendship Hotel in Beijing for one night, but she wants to know how the noise is inside the Friendship Hotel. In order to obtain accurate noise data as soon as possible, she prefers to find sensing users who are located inside the Friendship Hotel to complete the noise sensing task; for another example, Bob wants to find a restaurant in a scenic spot to eat; due to the restrictions of scenic spot tickets, geometric conditions, etc., he prefers to find users who are in the scenic area to help him determine the specific location of the restaurant. However, users uploading their own location information or querying the range may reveal their privacy. For example, Alice uploads the query information of Beijing Friendship Hotel to the mobile crowdsensing platform, and the adversary will infer Alice's economic situation based on this and then guess Alice's identity, occupation, family status, economic level, and other sensitive information [7]. Therefore, the geometric range-based task allocation requires strict preservation of users' location and query information.

Currently, there are a sequences of schemes proposed to achieve the privacy-preserving geometric range query [8–17]. Specifically, in [8–10, 16], researchers proposed the geometric range query schemes based on public-key searchable encryption. However, due to the uses of time-consuming encryption algorithms, such as the Paillier homomorphic encryption algorithm, bilinear mapping, etc., these schemes have large computational and communication overheads. In order to improve the efficiency of location query, some researchers have also proposed some schemes based on symmetric searchable encryption [11–15, 18]. Compared with the schemes based on public-key searchable encryption, these schemes use a trusted data center architecture. That is, these schemes use a trusted data center to encrypt the user's location information and distribute search keys to the search user (the data requester). However, these schemes based on the trusted data center have some inherent defects: (1) users send data to the trusted data center, and then the trusted data center encrypts and uploads it to the mobile crowdsensing platform, which will increase the system overhead and increase the risk of data leakage, and (2) the search users have the same search key, and once the search key is leaked, it will cause a serious threat to the user's data security. In addition, due to the limitations of the geometric environment and building structure, how to achieve location queries in an arbitrary geometric range is also a challenge. Most of the current schemes [8–10, 15–17] can only support range query for a specific geometric range, such as circle or rectangle. The location query of a specific geometric range will return a lot of meaningless results, add extra overhead to the system, and reduce the user experience. In addition, most of the above solutions are based on the fact that the mobile crowdsensing platform cannot collude with users. However, in fact, the mobile crowdsensing platform is usually composed of cloud service providers, and cloud service providers may register as legitimate users or collude with users to obtain user-related key information. Then, cloud service providers try to obtain and infer the private data of other users based on user-related key information.

In this chapter, we consider the collusion attack between cloud servers and users and propose a geometric range-based privacy-preserving task allocation (GPTA) that supports arbitrary geometric range queries. The contributions of this chapter are summarized as follows:

- Based on non-collusive dual servers, we propose a geometric range-based privacy-preserving task allocation scheme (GPTA-L) that supports arbitrary geometric range and does not rely on the trusted data center. GPTA-L utilizes matrix decomposition to generate encryption/re-encryption keys, utilizes the polynomial fitting technology to generate user location information and query range, and utilizes random matrix multiplication technology to preserve user location and query privacy. Taking advantage of the properties of the matrix, GPTA-L enables the mobile crowdsensing platform to find sensing users located within the geometric range without knowing the user's location and query information. At the same time, GPTA-L can resist the collusion attack of the single cloud server and user.
- In order to further improve the efficiency of location query, on the basis of GPTA-L, we study the historical query behaviors of data requesters and propose GPTA-F that can achieve nonlinear query efficiency. Specifically, GPTA-F first sets tags for the data requester's historical queries. Next, GPTA-F determines whether the query results under the tag are within the range of the new query by matching the new query with the tag.
- Security analysis proves that the proposed scheme can effectively preserve the user's location and query privacy. Experiments based on the mobile crowdsensing environment verify that the proposed scheme has high computational efficiency in data encryption and location query.

The rest of this chapter is organized as follows: Sect. 4.1.2 reviews the related works on task allocation based on the geometric range; Sect. 4.1.3 provides a brief description of polynomial fitting; Sect. 4.2 describes the architecture overview of GPTA from system model, security model, and design goals; Sect. 4.3 presents the detailed design of GPTA-L and GPTA-F; Sect. 4.4 gives the security analysis of GPTA; Sect. 4.5 presents the performance evaluation and analysis; and Sect. 4.6 summarizes the work of this chapter.

## 4.1.2 Related Works

Task allocation based on the geometric range is an important part of MCS. Liu et al. [9] proposed a privacy-preserving task allocation scheme based on a non-collusion two-server model by using Paillier homomorphic encryption algorithm and garbled circuits, which can help data requesters find sensing users located within a certain range. Zhai et al. [19] used the travel cost to represent the user's location information and designed an anonymous data aggregation protocol to prevent the perception platform from guessing the user's real location. Wang et al. [20] proposed a privacy-preserving task allocation scheme that selects users with the shortest travel distance based on differential privacy technology. Similarly, Wang et al. [7] also designed a personalized task allocation scheme using differential privacy technology. Specifically, this scheme requires each sensing user to upload the blurred distance and personal privacy budget, and by designing a probabilistic winner selection mechanism protocol, the scheme can assign tasks to appropriate sensing users. In addition, there are currently some privacy-preserving task allocation schemes that focus on other factors such as anonymity, reputation value, fog computing scenarios, etc. [21–24].

To the best of our knowledge, there is currently no privacy-preserving task allocation scheme that supports arbitrary geometric range queries. Therefore, this chapter will introduce the existing schemes that can realize privacy-preserving retrieval in any geometric range. Specifically, Zhu et al. [8] designed a privacy-preserving range retrieval using the BGN algorithm and hash table. However, their scheme has a lot of computational and communication overhead due to time-consuming cryptographic operations. To improve retrieval efficiency, Wang et al. proposed several privacy-preserving range retrieval schemes [12, 13, 26] using technologies, e.g., Shen-Shi-Waters (SSW) [27]. Among them, scheme [26] can only realize the circle range, and scheme [12] uses the Bloom filter to represent the user's location information and realizes the retrieval of arbitrary range by calculating the inner product of the two Bloom filters. However, as the range of the query increases, the Bloom filter will become larger and larger, which will introduce a high computational cost. Further, scheme [13] improves query efficiency and accuracy by transforming user location and query range into the form of equation vectors. Xu et al. [11] used polynomial fitting functions and secure kNN [28] technology to achieve efficient and secure arbitrary range queries with access control [29]. Wang et al. [14] implemented a privacy-preserving arbitrary geometric range query using a quadtree and SHVE [30] algorithm. Most of the above solutions use the architecture of a trusted data center, so it is difficult to apply in mobile crowdsensing scenarios. Table 4.1 shows the comparison of the functions and security properties of the proposed scheme and existing schemes.

**Table 4.1** Comparison between GPTA and other works in terms of function and security properties

| Scheme | No trusted data center | Query method | Nonlinear | Security | Efficiency |
|---|---|---|---|---|---|
| SRQC [8] | × | Circle | × | IND-CPA | High |
| ETA [9] | ✓ | Circle | × | IND-CPA | Low |
| EPPD [10] | ✓ | Circle | × | IND-CPA | High |
| EGRQ [11] | × | Arbitrary range | ✓ | IND-CLS-CPA | Very high |
| GRSE [12] | × | Arbitrary range | ✓ | IND-CPA | Low |
| FastGeo [13] | × | Arbitrary range | ✓ | IND-CPA | High |
| PBRQ-L [14] | × | Arbitrary range | × | IND-CPA | High |
| PPTA [25] | ✓ | Arbitrary range | × | IND-CPA | High |
| CRSE [26] | × | Circle | × | IND-CPA | Low |
| GPTA-L | ✓ | Arbitrary range | × | IND-CPA | High |
| GPTA-F | ✓ | Arbitrary range | ✓ | IND-CPA | Very high |

### *4.1.3  Preliminary*

In this section, we give a brief description of the polynomial fitting technology before presenting the details of our proposed GPTA.

#### 4.1.3.1  Polynomial Fitting

The polynomial fitting technology can fit the curve according to the coordinates of multiple points. As shown in Fig. 2.1, for the geometric range $(\theta_a, \theta_b)$, select several coordinate points on the curve $(\theta_a, \theta_b)$ to fit two curves $(\theta_a{}^*, \theta_b{}^*)$, so that the generated curve is consistent with the original geometric curve as much as possible. The fitted curve equation can be expressed as $\theta_a{}^*(x) = a_0 + a_1 x + a_2 x^2 + \cdots + a_n x^n$, $\theta_b{}^*(x) = b_0 + b_1 x + b_2 x^2 + \cdots + b_n x^n$, where $a_i, b_i (i \in [0, n])$ are the coefficients of the two curve equations, respectively, and $n$ is the highest power of the curve equation. Given geometric range $(\theta_a{}^*, \theta_b{}^*)$, judge whether a node $(x_i, y_i)$ is within the geometric range through the following steps:

- Step 1: Calculate $(\theta_a{}^*(x_i) - y_i)$ and judge whether the value is greater than 0. If it is greater than 0, proceed to step 2; otherwise, the node is not within the geometric range.
- Step 2: Calculate $(\theta_b{}^*(x_i) - y_i)$ and judge whether the value is less than 0. If it is less than 0, the node is within the geometric range; otherwise, the node is not within the geometric range.

The polynomial fitting technology is an approximate algorithm. The fitting function generated by this technology may not fit the given geometric range perfectly, resulting in certain fitting errors. In order to improve the fitting accuracy, the fitting error can be controlled in a very small range through technologies such as the orthogonal family. Interested readers can refer to papers [31, 32] for more technical details. In this chapter, the polynomial fitting technology is used to fit any geographical range and, based on the fitted curve, judge whether the locations of the sensing users are within the geographical range queried by the data requester.

## 4.2  Architecture Overview

In this section, we will describe the system model, security model, and design goals of GPTA.

### 4.2.1  System Model

As shown in Fig. 4.1, GPTA contains four entities, namely, trusted authority (TA), mobile crowdsensing platform (CP), data requester (DR), and sensing user (SU). The functions of each entity are defined as follows:

- Trusted authority: The trusted authority is responsible for generating system parameters for entities participating in task allocation (steps ① and ②). After the system is initialized, the trusted authority will stay offline.
- Mobile crowdsensing platform: The mobile crowdsensing platform is responsible for storing the location information of sensing users and provides task allocation services for data requesters (steps ⑤ and ⑥). In GPTA, two non-colluding cloud service providers, i.e., $S_A$ and $S_B$, such as Amazon Cloud and Microsoft Cloud, are used as the mobile crowdsensing platform.
- Data requester: Data requesters attempt to find users within their required geometric range without compromising their query privacy. Therefore, data requesters generate the query trapdoors and send trapdoors to the mobile crowdsensing platform (step ④).
- Sensing user: Sensing users usually send their locations to the mobile crowdsensing platform for accepting data collection tasks. To preserve their location privacy, sensing users encrypt the locations before sending the data to the mobile crowdsensing platform (step ③).

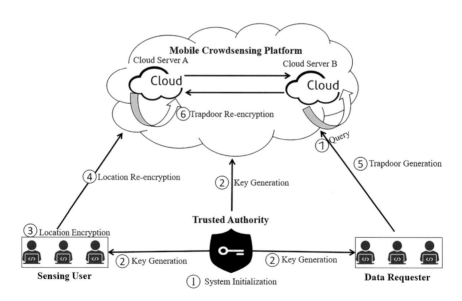

**Fig. 4.1**  System model of GPTA

## 4.2.2 Security Model

In the system model of GPTA, we assume that the trusted authority is fully trusted, and the communication channel between the trusted authority and any other entity is secure. Sensing users, data requesters, and the mobile crowdsensing platform are semi-trusted. That is, they will execute the protocol honestly but attempt to obtain private information from encrypted task information and trapdoors. Based on the adversary's attack capability, we consider the following two attacks:

- Passive attack: Passive attack means that the adversary can observe the ciphertexts but does not know the plaintexts corresponding to the ciphertexts. This attack is similar to a ciphertext-only attack in cryptography. It may be launched when an adversary monitors a user's communication channel or the mobile crowdsensing platform receives the user's ciphertext data.
- Active attack: In addition to the ciphertexts, an adversary with active attack capability may know the plaintexts corresponding to the ciphertexts. Specifically, an external adversary may learn the sensing users' plaintext-ciphertext pairs by observing the users' location information and the ciphertexts. This attack is similar to a known-plaintext attack in cryptography. Additionally, if a cloud server colludes with the sensing users or data requesters, the cloud server can even select some special plaintexts to obtain the corresponding ciphertexts. This attack is similar to a chosen-plaintext attack in cryptography.

We assume that the two cloud servers do not collude. That is, one cloud server will not share its own key information with the other cloud server or cooperate with the other cloud server to do behaviors that may compromise user privacy. This is a common assumption in many schemes based on the non-collusive server model [33–36]. In practical applications, a cloud server may collude with some users to obtain relevant key information. However, we assume that colluding users cannot upload a large number of locations or query requests. This assumption is reasonable because (1) a sensing user is only allowed to upload one or a limited amount of location information in the system and (2) data requesters usually need to pay for their query requests. In fact, since another cloud server and the sensing users will obtain rewards from the task allocations, we allow one cloud server to submit a query request to find sensing users within a certain search range. In addition, in this chapter, we mainly focus on privacy preservation for location and query. Although private information such as task content and user identity are also important, they are not considered in this chapter. In practical applications, users can encrypt task information [25] or use anonymous authentication technology, such as pseudonym [37], random anonymous authentication algorithm [38], one-way hash chain [35], etc., to hide their identity information. Since there are many studies [33–36] that can be used to effectively detect and resist the Sybil attack, the Sybil attack is also not considered in this chapter. In addition, we allow the cloud servers to obtain some information, including (1) the number of encrypted locations and query trapdoors, (2) the sensing user ID returned for a query, and (3) whether an encrypted location is used by two different trapdoors. These information are allowed to be leaked by default in most searchable encryption schemes [11–15].

### 4.2.3  Design Goals

According to the above system model and security model, GPTA should achieve the following privacy and functional goals:

- Data privacy: Privacy of location and query should be strictly preserved. That is, given a ciphertext, the adversary cannot recover the location or query information from it. It is noted that we allow the data requesters to determine the approximate locations of sensing users through the query since the data requesters have to pay for the query behaviors, and the cloud server and the sensing users will also obtain rewards accordingly.
- Query accuracy: GPTA should guarantee the accuracy of the query. That is, given a query with an arbitrary geometric range, GPTA should find sensing users within the range.
- Query efficiency: GPTA should guarantee the efficiency of the query. That is, given a query, GPTA should efficiently perform query operations over ciphertexts.

## 4.3  Detailed Design

In this section, we will describe GPTA in detail. Specifically, GPTA mainly includes seven steps: (1) system initialization, (2) key generation, (3) location encryption, (4) location re-encryption, (5) trapdoor generation, (6) trapdoor re-encryption, and (7) query. In particular, we first propose GPTA-L with linear query efficiency based on the polynomial fitting technology and random matrix multiplication technology and give a detailed correctness analysis of GPTA-L. Then, we propose GPTA-F with nonlinear query efficiency based on the designed geometric query structure and give a detailed correctness analysis of GPTA-F.

### 4.3.1  Proposed GPTA-L Scheme

**System Initialization**  Given the highest power $n$ of a polynomial fitting function, the trusted authority generates three $(n + 10) \times (n + 10)$-dimensional invertible random matrices $\{M_1, M_A, M_B\}$ as the master secret key.

**Key Generation**  Given a valid identity of sensing user $u_i$, the trusted authority first chooses a $(n + 10) \times (n + 10)$-dimensional invertible random matrix $A_i$ and generates a $(n + 10) \times (n + 10)$-dimensional lower triangular random matrix $I_i$ with main diagonal elements $(0, 0, 1, \cdots, 1, 1, 1, 1, 1, 0, 0)$ as follows:

$$
I_i =
\begin{bmatrix}
0 & 0 & 0 & 0 & \cdots & 0 & 0 & 0 \\
0 & 0 & 0 & 0 & \cdots & 0 & 0 & 0 \\
* & * & 1 & 0 & \cdots & 0 & 0 & 0 \\
* & * & * & 1 & 0 & \cdots & 0 & 0 \\
\multicolumn{8}{c}{\cdots\cdots\cdots\cdots\cdots\cdots\cdots\cdots} \\
* & \cdots\cdots & * & * & 1 & 0 & 0 \\
* & \cdots\cdots\cdots & * & 0 & 0 \\
* & \cdots\cdots\cdots & * & 0 & 0
\end{bmatrix}_{(n+10)\times(n+10)}
,
$$

where $*$ represents the random in the lower triangular random matrix. Next, the trusted authority computes $A_i' = M_1 \times I_i \times A_i^{-1}$. Given a valid identity of data requester $u_q$, the trusted authority first chooses a $(n + 10) \times (n + 10)$-dimensional invertible random matrix $B_q$ and generates a $(n + 10) \times (n + 10)$-dimensional lower triangular random matrix $S_q$ with similar construction to $I_i$ with main diagonal elements $(0, 0, 1, \cdots, 1, 1, 1, 1, 1, 0, 0)$. Next, the trusted authority computes $B_q' = B_q^{-1} \times S_q \times M_1^{-1}$. Then, the trusted authority generates a perturbation matrix $R_i = A_i \times I_i'$ for each sensing user and a perturbation matrix $R_q = S_q' \times B_q$ for each data requester, where $I_i'$ is a $(n + 10) \times (n + 10)$-dimensional lower triangular random matrix with similar construction to $I_i$ with main diagonal elements $(0, 0, 0, \cdots, 0, \gamma_{i,1}, \gamma_{i,2}, 1, 1, 0, 0)$, $S_q'$ is a $(n+10) \times (n+10)$-dimensional lower triangular random matrix with similar construction to $I_i$ with main diagonal elements $(0, 0, 0, \cdots, 0, \gamma_{q,1}, \gamma_{q,2}, 1, 1, 0, 0)$, and $\{\gamma_{i,1}, \gamma_{i,2}, \gamma_{q,1}, \gamma_{q,2}\} > 0$. Finally, the trusted authority utilizes the secure communication channel to send the encryption key $A_i$ to the sensing user $u_i$, send the encryption key $B_q$ to the data requester $u_q$, send the re-encryption key $(M_A, M_B, A_i', R_i, R_q)$ to the cloud $S_A$, and send the re-encryption key $M_B^{-1} \times B_q'$ to the cloud $S_B$.

**Location Encryption**  The sensing user $u_i$ first obtains its location coordinates $p_i = (x_i, y_i)$ and then performs the following operations:

1. $u_i$ extends $p_i$ to a $(n + 10)$-dimensional vector $l_i = (0, 0, 1, x_i, \cdots, x_i^n, y_i, 0, 0, 0, 0, 0, 0)$ and generates a lower triangular random matrix $P_i$ with similar construction to $I_i$ with main diagonal element $l_i$.
2. $u_i$ utilizes the encryption key $A_i$ to encrypt the location information as $C_i = A_i \times P_i$.
3. $u_i$ sends its ID and the encrypted $C_i$ to the cloud $S_A$.

**Location Re-encryption**  After receiving the ciphertext $C_i$ from $u_i$, $S_A$ first searches the re-encryption key $(M_A, A_i', R_i)$ and generates a $(n + 10) \times (n + 10)$-dimensional lower triangular random matrix $D_i$ with main diagonal elements $(0, 0, v_i, \cdots, v_i, 1, 1, r_{i,1}, r_{i,2}, 0, 0)$, where $v_i$, $r_{i,1}$, and $r_{i,2}$ are random positive numbers. Then, $S_A$ re-encrypts $C_i$ as follows:

$$
\begin{aligned}
\tilde{C}_i &= A'_i \times (C_i + R_i) \times D_i \times M_A \\
&= A'_i \times A_i \times (P_i + I'_i) \times D_i \times M_A \\
&= M_1 \times I_i \times (P_i + I'_i) \times D_i \times M_A.
\end{aligned}
\tag{4.1}
$$

At last, $S_A$ sends $\tilde{C}_i$ to the cloud $S_B$ and stores the random numbers $r_{i,1}$ and $r_{i,2}$.

**Trapdoor Generation**  Given a geometric range, the data requester $u_q$ generates a query trapdoor according to the following steps:

1. $u_q$ utilizes the polynomial fitting technology to generate two curve equations $\theta^*_a(x) = a_0 + a_1 x + a_2 x^2 + \cdots + a_n x^n$ and $\theta^*_b(x) = b_0 + b_1 x + b_2 x^2 + \cdots + b_n x^n$ for fitting the geometric range. Then, $u_q$ extracts the coefficients of $\theta^*_a(x)$ and $\theta^*_b(x)$ to generate two vectors $q_a = (a_0, a_1, \cdots, a_n)$ and $q_b = (b_0, b_1, \cdots, b_n)$.
2. $u_q$ extends $q_a$ and $q_b$ to two $(n+10)$-dimensional vectors $l_a = (0, 0, a_0, a_1, \cdots, a_n, -1, 0, 0, 0, 0, 0, 0)$ and $l_b = (0, 0, b_0, b_1, \cdots, b_n, -1, 0, 0, 0, 0, 0, 0)$, respectively. Then, $u_q$ generates two $(n + 10) \times (n + 10)$-dimensional lower triangular random matrices $Q_a$ and $Q_b$ with similar construction to $I_i$ with main diagonal elements $l_a$ and $l_b$, respectively. Next, $u_q$ utilizes the encryption key $B_q$ to encrypt $Q_a$ and $Q_b$ as $T_a = Q_a \times B_q$ and $T_b = Q_b \times B_q$.
3. $u_q$ sends $u_q$'s ID and the encrypted trapdoors $T_a$ and $T_b$ to the cloud $S_A$.

**Trapdoor Re-encryption**  After receiving the ciphertext $T_a$ and $T_b$ from $u_q$, $S_A$ first searches the re-encryption key $(M_A, M_B, R_q)$. Then, $S_A$ generates two $(n + 10) \times (n + 10)$-dimensional lower triangular random matrices $D_a$ and $D_b$ with main diagonal elements $(0, 0, v_a, \cdots, v_a, r_{a,1}, r_{a,2}, r_{q,1}, r_{q,2}, 0, 0)$ and $(0, 0, v_b, \cdots, v_b, r_{b,1}, r_{b,2}, r_{q,1}, r_{q,2}, 0, 0)$, respectively, where $v_a, v_b, r_{a,1}, r_{a,2}, r_{b,1}, r_{b,2}, r_{q,1}$, and $r_{q,2}$ are random positive numbers. Then, $S_A$ re-encrypts $T_a$ and $T_b$ as follows:

$$
\begin{aligned}
\tilde{T}_a &= M_A^{-1} \times D_a \times (T_a + R_q) \times M_B \\
&= M_A^{-1} \times D_a \times (Q_a \times B_q + S'_q \times B_q) \times M_B \\
&= M_A^{-1} \times D_a \times (Q_a + S'_q) \times B_q \times M_B,
\end{aligned}
\tag{4.2}
$$

$$
\begin{aligned}
\tilde{T}_b &= M_A^{-1} \times D_b \times (T_b + R_q) \times M_B \\
&= M_A^{-1} \times D_b \times (Q_b \times B_q + S'_q \times B_q) \times M_B \\
&= M_A^{-1} \times D_b \times (Q_b + S'_q) \times B_q \times M_B.
\end{aligned}
\tag{4.3}
$$

Next, $S_A$ sends $\tilde{T}_a$, $\tilde{T}_b$, and $\{r_{i,1} r_{q,1} + r_{i,2} r_{q,2}\}_{i=1}^m$ to the cloud $S_B$, where $m$ represents the total number of location ciphertexts. Finally, $S_B$ searches the re-encryption key $M_B^{-1} \times B_q'$ to perform the following operations:

$$\bar{T}_a = \tilde{T}_a \times M_B^{-1} \times B_q'$$

$$= M_A^{-1} \times D_a \times (Q_a + S_q') \times B_q \times M_B \times M_B^{-1} \times B_q' \tag{4.4}$$

$$= M_A^{-1} \times D_a \times (Q_a + S_q') \times S_q \times M_1^{-1},$$

$$\bar{T}_b = \tilde{T}_b \times M_B^{-1} \times B_q'$$

$$= M_A^{-1} \times D_b \times (Q_b + S_q') \times B_q \times M_B \times M_B^{-1} \times B_q' \tag{4.5}$$

$$= M_A^{-1} \times D_b \times (Q_b + S_q') \times S_q \times M_1^{-1}.$$

**Query** Given a re-encrypted location information $\tilde{C}_i$ and re-encrypted trapdoors $(\bar{T}_a, \bar{T}_b)$, the cloud $S_B$ performs the following operations:

$$\Delta_a = tr(\tilde{C}_i \times \bar{T}_a) - (r_{i,1}r_{q,1} + r_{i,2}r_{q,2})$$

$$= tr(M_1 \times I_i \times (P_i + I_i') \times D_i \times M_A \times M_A^{-1} \times D_a \times (Q_a + S_q')$$

$$\times S_q \times M_1^{-1}) \tag{4.6}$$

$$- (r_{i,1}r_{q,1} + r_{i,2}r_{q,2})$$

$$= v_i v_a(\theta_a^*(x_i) - y_i) + r_{a,1}\gamma_{q,1}\gamma_{i,1} + r_{a,2}\gamma_{q,2}\gamma_{i,2},$$

$$\Delta_b = tr(\tilde{C}_i \times \bar{T}_b) - (r_{i,1}r_{q,1} + r_{i,2}r_{q,2})$$

$$= tr(M_1 \times I_i \times (P_i + I_i') \times D_i \times M_A \times M_A^{-1} \times D_b \times (Q_b + S_q')$$

$$\times S_q \times M_1^{-1}) \tag{4.7}$$

$$- (r_{i,1}r_{q,1} + r_{i,2}r_{q,2})$$

$$= v_i v_b(\theta_b^*(x_i) - y_i) + r_{b,1}\gamma_{q,1}\gamma_{i,1} + r_{b,2}\gamma_{q,2}\gamma_{i,2},$$

where $tr(\cdot)$ represents the matrix trace operation. In order to eliminate the influence of $r_{a,1}\gamma_{q,1}\gamma_{i,1} + r_{a,2}\gamma_{q,2}\gamma_{i,2}$ and $r_{b,1}\gamma_{q,1}\gamma_{i,1} + r_{b,2}\gamma_{q,2}\gamma_{i,2}$, we need to control the value range of these random numbers. Specifically, if we assume that $|\theta_a^*(x_i) - y_i|$ and $|\theta_b^*(x_i) - y_i|$ are only greater than a very small positive number $\mathcal{A}$, and $\{r_{a,1}, r_{a,2}, r_{b,1}, r_{b,2}, \gamma_{i,2}, \gamma_{i,2}, \gamma_{q,1}, \gamma_{q,2}\}$ is smaller than a positive number $B$, then $\{v_i, v_a, v_b\} \gg \sqrt{\frac{2B^3}{A}}$ should be satisfied. Thus, if $\Delta_a > 0$ and $\Delta_b < 0$ are satisfied simultaneously, it can be known that $R_{i,q} = 1$. That is, the sensing user $u_i$ locates within the geometric range of the data requester $u_q$. Then, $S_B$ adds the ID of $u_i$ into the result list $RS_q$. Otherwise, $S_B$ selects another sensing user's encrypted data to repeat the above operations. After querying all encrypted data, $S_B$ can get all sensing users $RS_q = \{u_i\}_{R_{iq}=1}$ located within the geometric range of the data requester $u_q$.

### 4.3.2  Correctness of GPTA-L

In this section, we will give the correct analysis of GPTA-L. Firstly, it is noted that the lower triangular random matrix means that the elements above the main diagonal are all 0, and the trace of a matrix equals the sum of the main diagonal elements. According to **Lemma 3.1** and **Lemma 3.2** in Chap. 3, we know that (1) given an invertible matrix $M$ and a square matrix $P$ with the same dimension, $tr(M \times P \times M^{-1}) = tr(P)$, and (2) given two lower triangular matrices $P$ and $S$ with the same dimension, if their main diagonal elements are vectors $\mathbf{P}$ and $\mathbf{S}$, respectively, then $tr(P \times S) = \mathbf{P} \circ \mathbf{S}$, where $\mathbf{P} \circ \mathbf{S}$ represents the inner product of $\mathbf{P}$ and $\mathbf{S}$. Based on this, we have the following theorems.

**Theorem 4.1** *Taking the vectors $l_i$, $l_a$, and $l_b$ as inputs, we can obtain the cipher-texts $\tilde{C}_i$, $\bar{T}_a$, $\bar{T}_b$ in GPTA-L. If $l_i \circ l_a > 0(< 0) \Leftrightarrow tr(\tilde{C}_i \times \bar{T}_a) - (r_{i,1}r_{q,1} + r_{i,2}r_{q,2}) > 0(< 0)$ and $l_i \circ l_b > 0(< 0) \Leftrightarrow tr(\tilde{C}_i \times \bar{T}_b) - (r_{i,1}r_{q,1} + r_{i,2}r_{q,2}) > 0(< 0)$, then the correctness of GPTA-L is proven.*

***Proof*** Based on the design of GPTA-L, we first extend the vector $l_i$ to a lower triangular random matrix $P_i$ with the main diagonal element $l_i$. Next, $P_i$ is encrypted as $C_i$ on the sensing user side and is re-encrypted as $\tilde{C}_i$ on the cloud side. Then, the query vectors $l_a$ and $l_b$ are extended to two lower triangular random matrices $T_a$ and $T_b$ whose main diagonal elements are $l_a$ and $l_b$, respectively. Next, $T_a$ and $T_b$ are re-encrypted as $\bar{T}_a$ and $\bar{T}_b$, respectively. $S_B$ computes $\Delta_a = tr(\tilde{C}_i \times \bar{T}_a) - (r_{i,1}r_{q,1} + r_{i,2}r_{q,2})$ and $\Delta_b = tr(\tilde{C}_i \times \bar{T}_b) - (r_{i,1}r_{q,1} + r_{i,2}r_{q,2})$. We first take $\Delta_a$ as an example to analyze the correctness.

**Step 1:** Since $M_1$ is an invertible matrix, it can be known that $\Delta_a = tr(M_1 \times I_i \times (P_i + I'_i) \times D_i \times M_A \times M_A^{-1} \times D_a \times (Q_a + S'_q) \times S_q \times M_1^{-1} - (r_{i,1}r_{q,1} + r_{i,2}r_{q,2}) = tr(I_i \times (P_i + I'_i) \times D_i \times M_A \times M_A^{-1} \times D_a \times (Q_a + S'_q) \times S_q - (r_{i,1}r_{q,1} + r_{i,2}r_{q,2})$ based on **Lemma** 3.1 in the Sect. 3.3.2.

**Step 2:** Since $I_i$, $P_i$, $I'_i$, $D_i$, $D_a$, $Q_a$, $S'_q$, and $S_q$ are lower triangular random matrices, whose main diagonal elements are $\mathbf{I_i} = (0, 0, 1, \cdots, 1, 1, 1, 1, 1, 0, 0)$, $l_i$, $\mathbf{I'_i} = (0, \cdots, 0, \gamma_{i,1}, \gamma_{i,2}, 1, 1, 0, 0)$, $\mathbf{D_i} = (0, 0, v_i, \cdots, v_i, 1, 1, r_{i,1}, r_{i,2}, 0, 0)$, $\mathbf{D_a} = (0, 0, v_a, \cdots, v_a, r_{a,1}, r_{a,2}, r_{q,1}, r_{q,2}, 0, 0)$, $l_a$, $\mathbf{S'_q} = (0, \cdots, 0, \gamma_{q,1}, \gamma_{q,2}, 1, 1, 0, 0)$, and $\mathbf{S_q} = (0, 0, 1, \cdots, 1, 1, 1, 1, 1, 0, 0)$, respectively, based on **Lemma** 3.2 in the Sect. 3.3.2, it can be known $\Delta_a = \mathbf{I_i} \circ (l_i + \mathbf{I'_i}) \circ \mathbf{D_i} \circ \mathbf{D_a} \circ (l_a + \mathbf{S'_q}) \circ \mathbf{S_q} - (r_{i,1}r_{q,1} + r_{i,2}r_{q,2})$. That is, $\Delta_a = v_i v_a (\theta_a^*(x_i) - y_i) + r_{a,1}\gamma_{q,1}\gamma_{i,1} + r_{a,2}\gamma_{q,2}\gamma_{i,2}$.

**Step 3:** Since we assume that $\{v_i, v_a\} \gg \sqrt{\frac{2B^3}{A}}$, $|\theta_a^*(x_i) - y_i| > A$, and $0 < \{r_{a,1}, r_{a,2}, r_{b,1}, r_{b,2}, \gamma_{i,1}, \gamma_{i,2}, \gamma_{q,1}, \gamma_{q,2}\} < B$, it can be known that $\theta_a^*(x_i) - y_i >$

$0 \rightarrow \Delta_a > 0$ and $\theta_a^*(x_i) - y_i < 0 \rightarrow \Delta_a < 0$. In addition, it can be known that $\Delta_a > 0 \rightarrow \theta_a^*(x_i) - y_i > 0$ and $\Delta_a < 0 \rightarrow \theta_a^*(x_i) - y_i < 0$.

Based on similar correctness analysis process, it can be known that $\theta_b^*(x_i) - y_i > 0 \rightarrow \Delta_b > 0$ and $\theta_b^*(x_i) - y_i < 0 \rightarrow \Delta_b < 0$. In addition, it can be known that $\Delta_b > 0 \rightarrow \theta_b^*(x_i) - y_i > 0$ and $\Delta_b < 0 \rightarrow \theta_b^*(x_i) - y_i < 0$. Thus, Theorem 4.1 is proven.

### 4.3.3 Proposed GPTA-F Scheme

Although GPTA-L can effectively find sensing users in arbitrary geometric ranges, it still faces the problem of query efficiency. As described in GPTA-L, $S_B$ needs to perform the matching calculation for each sensing user's encrypted data. When the amount of data is large, GPTA-L with a linear query will greatly limit the efficiency of sensing user query. In order to improve the sensing user query efficiency, we design an index structure based on the historical query. Based on this, we propose GPTA-F to reduce user query overhead. GPTA-F is inspired by the following: (1) the location of the queried sensing user is within the range of the query and remains unchanged within a certain time range, and (2) if there is an intersection between the two query ranges, then sensing users within one range may be in another range. Otherwise, if there is no intersection between the two query ranges, sensing users in one range cannot be in the other. Therefore, the query range of the data requester can be regarded as a parent node, and the query result can be regarded as a child node under the parent node. Given a new geometric range query, the user query phase for GPTA-F is as follows:

- Step 1. Randomly pick a parent node, and check whether there is an intersection between the new query and the parent node. If there is an intersection, go to step 2; otherwise, repeat step 1.
- Step 2. Randomly select a child node from the current parent node, and check whether the child nodes belong to the new query range. If so, add the sensing user to the result list; otherwise, repeat step 2.
- Step 3. For the sensing users who have not joined the parent node, check whether the sensing users belong to the new query range. If so, add the sensing users to the result list and mark the new query as the parent node of the sensing user; otherwise, repeat step 3.

However, with the arbitrary property of the query range, it is difficult to detect whether there is an intersection between the new query range and the historical query range. In order to address this problem, our strategy is to generate a circle outside the query range to cover the query range, as shown in Figs. 4.2, 4.3, 4.4, and 4.5. Then, we utilize the relationship between circles, as shown in Figs. 4.6, 4.7, 4.8, and 4.9, to determine whether there is an intersection between the two queries. Specifically, given two circles, the relationship between the two query ranges can be

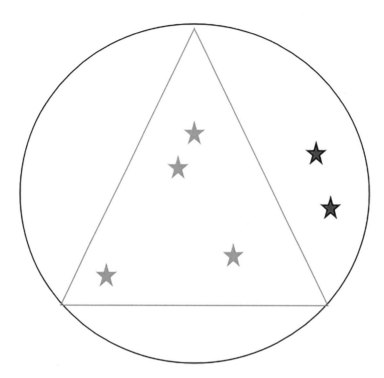

**Fig. 4.2** Generated circles covering the triangle range (orange represents the query range, black represents the circle covering the query)

judged by calculating the sum of the center-to-center distance and the radius of the two circles. That is, if the two circles intersect with each other, the two query ranges may intersect, and if the two circles are tangent or separated, the two query ranges cannot intersect. Therefore, we regard the newly generated circle as the parent node and the query results inside the circle as the child nodes.

In order to preserve the query privacy, the parent node (i.e., the circle) should be encrypted and then sent to the cloud server. Obviously, we can also utilize the encryption method in GPTA-L to preserve the privacy of the parent node. Therefore, the key in GPTA-F is how to judge whether two parent nodes intersect in ciphertexts. Specifically, to address the challenge, $u_q$ additionally generates a circle $C_q((x_q, y_q), RA_q)$ to cover the query range in the trapdoor generation phase, where $(x_q, y_q)$ represents the center of the circle and $RA_q$ represents the radius of the circle. According to $C_q((x_q, y_q), RA_q)$, $u_q$ generates two $(n+10) \times (n+10)$-dimensional lower triangular random matrices $C_{q,1}$ and $C_{q,2}$ with similar construction to $I_i$ with main diagonal elements $(0, 0, x_q, y_q, RA_q, (x_q^2 + y_q^2 - RA_q^2), 1, 0, \cdots, 0, 0)$ and $(0, 0, -2x_q, -2y_q, -2RA_q, 1, (x_q^2 + y_q^2 - RA_q^2), 0, \cdots, 0)$, respectively. Next, by location encryption phase, location re-encryption phase, trapdoor generation phase, and trapdoor re-encryption phase,

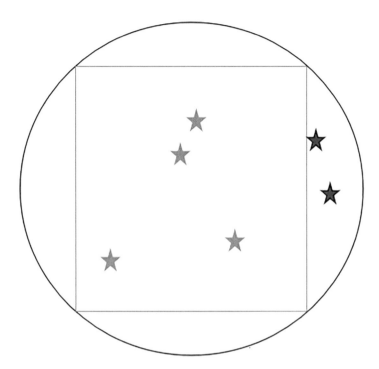

**Fig. 4.3** Generated circles covering the rectangle range (orange represents the query range, black represents the circle covering the query)

the cloud $S_B$ can obtain $\tilde{E}_{q,1} = M_1 \times I_q \times (C_{q,1} + I'_q) \times D_{q,1} \times M_A$ and $\bar{E}_{q,2} = M_A^{-1} \times D_{q,2} \times (C_{q,2} + S'_q) \times S_q \times M_1^{-1}$, where $\tilde{E}_{q,1}$ and $\bar{E}_{q,2}$ represent the parent node and trapdoor, respectively. It is noted that the $(n + 7)$-th and $(n + 8)$-th elements of the main diagonal elements in $D_{q,1}, D_{q,2}$ are set to 0 instead of random numbers. In the query phase, the sensing users in the query result $RS_q$ are regarded as the child nodes of $\tilde{E}_{q,1}$. Given the trapdoor $(\bar{E}_{k,2}, (\bar{T}_a, \bar{T}_b))$ from a data requester $u_k$, $S_B$ first chooses a parent node, such as $\tilde{E}_{q,1}$, and judge whether the two circles intersect by computing

$$
\begin{aligned}
ComS_{q,k} &= tr(\tilde{E}_{q,1} \times \bar{E}_{k,2}) \\
&= tr(M_1 \times I_q \times (C_{q,1} + I'_q) \\
&\quad \times D_{q,1} \times M_A \times M_A^{-1} \times D_{k,2} \times (C_{k,2} + S'_k) \times S_k \times M_1^{-1}) \quad (4.8) \\
&= v_q v_k ((x_k - x_q)^2 + (y_k - y_q)^2 - (RA_k + RA_q)^2) \\
&\quad + r_{k,1} \gamma_{q,1} \gamma_{k,1} + r_{k,2} \gamma_{q,2} \gamma_{k,2}.
\end{aligned}
$$

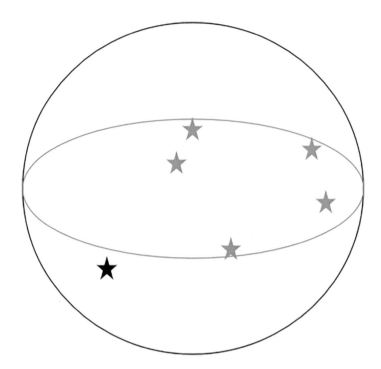

**Fig. 4.4** Generated circles covering the oval range (orange represents the query range, black represents the circle covering the query)

In order to eliminate the influence of $r_{k,1}\gamma_{q,1}\gamma_{k,1} + r_{k,2}\gamma_{q,2}\gamma_{k,2}$, we assume that $|(x_k - x_q)^2 + (y_k - y_q)^2 - (RA_k + RA_q)^2)|$ is greater than a very small positive number $C$, and $\{r_{k,1}, r_{k,2}, \gamma_{q,1}, \gamma_{q,2}, \gamma_{k,1}, \gamma_{k,2}\}$ is greater than 0 and smaller than a positive number $D$, then $\{v_q, v_k\} \gg \sqrt{\frac{2D^3}{C}}$ should be satisfied. Thus, if $ComS_{q,k} < 0$, $S_B$ chooses the child nodes within $\tilde{E}_{q,1}$ and computes $(\Delta_a, \Delta_b)$ to judge whether the child nodes locate within the query range. Otherwise, $S_B$ chooses another parent node $\tilde{E}_{q' \neq q}$ and continues to compute $ComS_{q',k}$. The detailed description of GPTA-F is shown in Fig. 4.10.

### 4.3.4  Correctness of GPTA-F

**Theorem 4.2** *Taking the vectors $l_i$, $l_a$, and $l_b$ as inputs, we can obtain the ciphertexts $\tilde{C}_i$, $\bar{T}_a$, $\bar{T}_b$ in GPTA-F. If $l_i \circ l_a > 0(< 0) \Leftrightarrow tr(\tilde{C}_i \times \bar{T}_a) - (r_{i,1}r_{q,1} + r_{i,2}r_{q,2}) > 0(< 0)$ and $l_i \circ l_b > 0(< 0) \Leftrightarrow tr(\tilde{C}_i \times \bar{T}_b) - (r_{i,1}r_{q,1} + r_{i,2}r_{q,2}) > 0(< 0)$, then the correctness of GPTA-F is proven.*

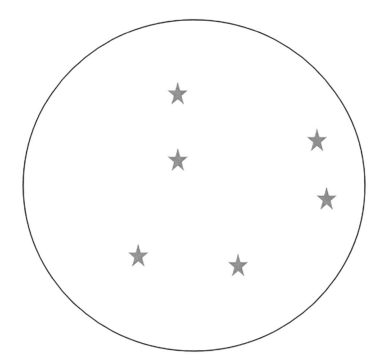

**Fig. 4.5** Generated circles covering the circle range (orange represents the query range, black represents the circle covering the query)

***Proof*** As shown in Fig. 4.7, it can be seen that if two circles intersect, the node located in one circle may be in the other circle; if the two circles are tangent or separated, the node located in one circle cannot be in the other circle. Based on this principle, in GPTA-F, we first check whether there is an intersection between the two circles by calculating $ComS_{q,k}$ and if $ComS_{q,k} < 0$, further check whether the nodes inside the circle are within the query range. Therefore, in order to prove the correctness of GPTA-F, the correctness of the $ComS_{q,k}$ calculation should be proved. That is, if and only if there is an intersection between the two circles $C_q((x_q, y_q), RA_q)$ and $C_k((x_k, y_k), RA_k)$, then $ComS_{q,k} < 0$ holds. The specific proof is provided in **Theorem 4.3**.

**Theorem 4.3** *In GPTA-F, if and only if there is an intersection between the two circles* $C_q((x_q, y_q), RA_q)$ *and* $C_k((x_k, y_k), RA_k)$, *then* $ComS_{q,k} < 0$ *holds.*

***Proof*** According to geometric properties, if there is an intersection between the two circles $C_q((x_q, y_q), RA_q)$ and $C_k((x_k, y_k), RA_k)$, then the distance between the centers of the two circles should be less than the sum of the radius of the two circles. That is, $(x_q - x_k)^2 + (y_q - y_k)^2 < (RA_k + RA_q)^2$. Recall GPTA-L and the proof of **Theorem 4.1**; it can be known that $ComS_{q,k} = v_q v_k((x_k - x_q)^2 + (y_k - y_q)^2 - (RA_k + RA_q)^2) + r_{k,1}\gamma_{q,1}\gamma_{k,1} + r_{k,2}\gamma_{q,2}\gamma_{k,2}$. Since we assume that $|(x_q -$

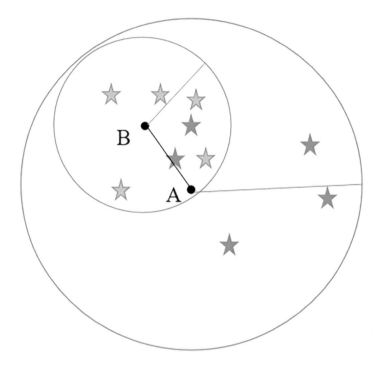

**Fig. 4.6** Relationship between two circles (include)

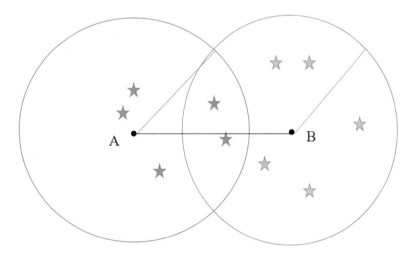

**Fig. 4.7** Relationship between two circles (intersect)

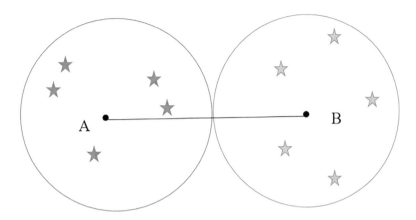

**Fig. 4.8** Relationship between two circles (outer)

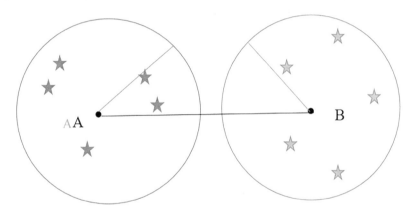

**Fig. 4.9** Relationship between two circles (outward)

$x_k)^2 + (y_q - y_k)^2 - (RA_k + RA_q)^2| > C, 0 < \{r_{k,1}, r_{k,2}, \gamma_{k,1}, \gamma_{k,2}, \gamma_{q,1}, \gamma_{q,2}\} < D$, and $\{v_q, v_k\} \gg \sqrt{\frac{2D^3}{C}}$, it can be known that if $(x_q - x_k)^2 + (y_q - y_k)^2 < (RA_k + RA_q)^2$, then $ComS_{q,k} < 0$, and if $(x_q - x_k)^2 + (y_q - y_k)^2 > (RA_k + RA_q)^2$, then $ComS_{q,k} > 0$.

In addition, we can see that if $ComS_{q,k} > 0$, then $(x_q - x_k)^2 + (y_q - y_k)^2 > (RA_k + RA_q)^2$, and if $ComS_{q,k} < 0$, then $(x_q - x_k)^2 + (y_q - y_k)^2 < (RA_k + RA_q)^2$. Thus, **Theorem 4.3** is proven.

**System Initialization:** Given the highest power $n$ of a polynomial fitting function, the trusted authority generates three $(n + 10) \times (n + 10)$-dimensional invertible random matrices $\{M_1, M_A, M_B\}$ as the master secret key.

**User Registration:** Given a valid identity of sensing user $u_i$, the trusted authority generates the encryption key $A_i$ and the re-encryption key $(M_A, A'_i, R_i)$, where $A'_i = M_1 \times I_i \times A_i^{-1}$ and $R_i = A_i \times I'_i$. Given a valid identity of data requester $u_q$, the trusted authority generates the encryption key $(A_q, B_q)$ and the re-encryption key $(A'_q, R_{q,1}, M_A, B'_q, R_{q,2}, M_B, M_B^{-1} \times B_q)$, where $A'_q = M_1 \times I_q \times A_q^{-1}$, $B'_q = B_q^{-1} \times S_q \times M_1^{-1}$, $R_{q,1} = A_q \times S'_q$, and $R_{q,2} = S'_q \times B_q$.

**Location Encryption:** The sensing user $u_i$ first obtains its location vector $l_i$, $u_i$ generates a matrix $P_i$ and encrypts $P_i$ as $C_i = A_i \times P_i$.

**Location Re-encryption:** After receiving the ciphertext $C_i$ from $u_i$, $S_A$ re-encrypts $C_i$ as $\tilde{C}_i = A'_i \times (C_i + R_i) \times D_i \times M_A$.

**Trapdoor Generation:** Given a geometric range $l_a$ and $l_b$, the data requester $u_q$ generates two matrices $Q_a$ and $Q_b$. Next, $u_q$ generates trapdoors $T_a = Q_a \times B_q$ and $T_b = Q_b \times B_q$. Then, assume that $C((x_q, y_q), RA_q)$ represents the circle recovering the query range, $u_q$ generates two matrices $C_{q,1}$ and $C_{q,2}$ and computes $E_{q,1} = A_q \times C_{q,1}$ and $E_{q,2} = C_{q,2} \times B_q$. Finally, $u_q$ sends $(u_q, (T_a, T_b), E_{q,1}, E_{q,2})$ to $S_A$.

**Trapdoor Re-encryption:** After receiving the trapdoors, $S_A$ computes $\{r_{i,1}r_{q,1} + r_{i,2}r_{q,2}\}_{i=1}^m$. Then, $S_A$ and $S_B$ search the re-encryption key and compute

$$
\begin{aligned}
S_A &: \tilde{T}_a = M_A^{-1} \times D_a \times (T_a + R_{q,2}) \times M_B, \\
S_B &: \bar{T}_a = \tilde{T}_a \times M_B^{-1} \times B'_q, \\
S_A &: \tilde{T}_b = M_A^{-1} \times D_b \times (T_b + R_q) \times M_B, \\
S_B &: \bar{T}_b = \tilde{T}_b \times M_B^{-1} \times B'_q, \\
S_A &: \tilde{E}_{q,1} = A'_q \times (E_{q,1} + R_{q,1}) \times D_{q,1} \times M_A, \\
S_A &: \tilde{E}_{q,2} = M_A^{-1} \times D_{q,2} \times (E_{q,2} + R_{q,2}) \times M_B, \\
S_B &: \bar{E}_{q,2} = \tilde{E}_{q,2} \times M_B^{-1} \times B'_q.
\end{aligned}
\tag{4.9}
$$

**Query:** $S_B$ performs query operations as follows:

- Step 1: For each parent node $\tilde{E}_{q,1}(q \neq k)$, $S_B$ computes

$$
ComS_{q,k} = tr(\tilde{E}_{q,1} \times \bar{E}_{k,2}).
\tag{4.10}
$$

If $ComS_{q,k} < 0$, $S_B$ performs step 2. Otherwise, $S_B$ repeats to perform step 1.
- Step 2: $S_B$ queries the child node of $\tilde{E}_{q,1}$ by computing

$$
\begin{aligned}
\Delta_a &= tr(\tilde{C}_i \times \bar{T}_a) - (r_{i,1}r_{q,1} + r_{i,2}r_{q,2}), \\
\Delta_b &= tr(\tilde{C}_i \times \bar{T}_b) - (r_{i,1}r_{q,1} + r_{i,2}r_{q,2}).
\end{aligned}
\tag{4.11}
$$

If $\Delta_a > 0$ and $\Delta_b < 0$, $S_B$ add the sensing user into the result list. Otherwise, $S_B$ chooses another child node to repeat step 2.
- Step 3: For a sensing user $u_l$ who does not locate within any parent node, $S_B$ computes

$$
\begin{aligned}
\Delta_a &= tr(\tilde{C}_i \times \bar{T}_a) - (r_{i,1}r_{q,1} + r_{i,2}r_{q,2}), \\
\Delta_b &= tr(\tilde{C}_i \times \bar{T}_b) - (r_{i,1}r_{q,1} + r_{i,2}r_{q,2}).
\end{aligned}
\tag{4.12}
$$

If $\Delta_a > 0$ and $\Delta_b < 0$, $S_B$ add the sensing user into the result list and set $\tilde{E}_{q,1}$ as the parent node of the current sensing user. Otherwise, $S_B$ chooses another child node who does not locate within any parent node to repeat step 3.

Finally, $S_B$ returns all sensing users within the result list to the data requester.

**Fig. 4.10** Detail in GPTA-F

## 4.4  Security Analysis

In this section, we will give the security analysis of GPTA. Since GPTA-L and GPTA-F mainly include three main steps: data encryption (location encryption, trapdoor generation), data re-encryption (location re-encryption, trapdoor re-encryption), and data query (user query), we will first discuss the security of each step separately. In addition, since the encryption and re-encryption operations of the location data and the query range are basically similar, we will take the location data as an example. For the convenience of description, we use GPTA to represent schemes GPTA-L and GPTA-F.

### *4.4.1  Security Analysis of the Data Encryption Phase*

#### 4.4.1.1  Security Under Passive Attack

Under the passive attack, the adversary may obtain a series of ciphertexts but does not know the corresponding plaintexts. Therefore, we first simulate the game $Passive_{\mathcal{A},GPTA}$ between the adversary $\mathcal{A}$ and the challenger $C$, as shown in Fig. 4.11. Next, we define the security of GPTA under the passive attack.

**Definition 4.1**  For any adversary $\mathcal{A}$ with probabilistic polynomial time (P.P.T), it has a negligible advantage in guessing $b' = b$ as follows:

$$|Pr(Passive_{A,GPTA}(\lambda) = 1) - \frac{1}{2}| \leq negl(\lambda), \tag{4.13}$$

and then GPTA can resist the passive attack.

**Theorem 4.4**  *GPTA can resist the passive attack in the data location phase.*

---

1.**Init:** Given a security parameter $\lambda$, the adversary $\mathcal{A}$ generates two databases $DB_0 = (p_{0,1}, p_{0,2}, \cdots, p_{0,t})$ and $DB_1 = (p_{1,1}, p_{1,2}, \cdots, p_{1,t})$, where $p_{i,j}(i \in 0, 1, j \in \{1, 2, \cdots, t\})$ represents the location data and each element $p_i, j$ ranges from 1 to $2^{\lambda}$. Then, $\mathcal{A}$ sends $DB_0$ and $DB_1$ to the challenger $C$.
2.**Setup:** The challenger $C$ performs the system initialization phase and user registration in GPTA to generate the master secret key and encryption key.
3.**Challenge:** According to the data received from **Init**, the challenger $C$ tosses a coin to choose $b \in \{0, 1\}$, and extends $p_{i,j}$ in $DB_b$ to a $(n + 10) \times (n + 10)$-dimensional lower triangular random matrix. Next, the challenger $C$ performs the location encryption phase to obtain $C_{b,j}$. Then, $C$ returns $EDB_b \leftarrow \{C_{b,j}\}_{j=1}^{t}$ to the adversary $\mathcal{A}$.
4.**Guess:** The adversary $\mathcal{A}$ guesses that $b'$ is either 0 or 1.
5.**Output:** If $b' = b$, output 1. Otherwise, output 0.

---

**Fig. 4.11** Passive attack $Passive_{\mathcal{A},GPTA}$

1.**Init:** Given a security parameter $\lambda$, the adversary $\mathcal{A}$ generates two databases $DB_0 = (p_{0,1}, p_{0,2}, \cdots, p_{0,t})$ and $DB_1 = (p_{1,1}, p_{1,2}, \cdots, p_{1,t})$, where $p_{i,j}(i \in \{0,1\}, j \in \{1, 2, \cdots, t\})$ represents the location data and each element $p_i, j$ ranges from 1 to $2^\lambda$. Then, $\mathcal{A}$ sends $DB_0$ and $DB_1$ to the challenger $C$.

2.**Setup:** The challenger $C$ performs the system initialization phase and the user registration phase in GPTA to generate the master secret key and encryption key.

3.**Phase 1:** In the $l$th query, the adversary $\mathcal{A}$ sends $DB_l = (p_{l,1}, p_{l,2}, \cdots, p_{l,t})$ to the challenger $C$. Next, $C$ extends the data in $DB_l$ to a $(n+10) \times (n+10)$-dimensional lower triangular random matrix. Then, $C$ performs the location encryption phase to obtain $C_{l,j}(j \in \{1, 2, \cdots, t\})$ and returns $C_{l,j}(j \in \{1, 2, \cdots, t\})$ to $\mathcal{A}$.

4.**Challenge:** Based on the received data in **Init**, the challenger $C$ tosses a coin to choose $b \in 0, 1$, and extends the data in $DB_b$ to a $(n+10) \times (n+10)$-dimensional lower triangular random matrix. Next, the challenger $C$ performs the location encryption phase to obtain $C_{b,j}(j \in \{1, 2, \cdots, t\})$ and returns $C_{b,j}(j \in \{1, 2, \cdots, t\})$ to the adversary $\mathcal{A}$.

5.**Phase 2:** The adversary $\mathcal{A}$ chooses another plaintext and repeats the **Phase 1**.

6.**Guess:** The adversary $\mathcal{A}$ guesses that $b'$ is either 0 or 1.

7.**Output:** If $b' = b$, output 1. Otherwise, output 0.

**Fig. 4.12** Active attack $Active_{\mathcal{A},GPTA}$

*Proof* According to the security model, an active adversary can obtain more information than a passive adversary. Therefore, if GPTA can resist the active attacks, then GPTA can resist the passive attacks. Therefore, in this section, we will omit specific proofs for passive attacks, and full proofs will be given in active attacks.

### 4.4.1.2  Security Under Active Attack

In addition to the ciphertexts, an active adversary can also obtain the plaintexts corresponding to the ciphertexts. In order to prove that GPTA can resist the chosen-plaintext attack, we first give $Active_{A,GPTA}$, as shown in Fig. 4.12, to simulate the game between the adversary $\mathcal{A}$ and the challenger $C$.

**Definition 4.2** For any adversary $\mathcal{A}$ with probabilistic polynomial time (P.P.T), it has a negligible advantage in guessing $b' = b$ as follows:

$$|Pr(Active_{A,GPTA}(\lambda) = 1) - \frac{1}{2}| \leq negl(\lambda), \tag{4.14}$$

and then GPTA can resist the active attack.

**Theorem 4.5** *GPTA can resist the active attack in the data encryption phase.*

*Proof* As shown in Fig. 4.6, we should prove that the adversary $\mathcal{A}$ cannot distinguish between $C_{0,j}$ and $C_{1,j}$. $\mathcal{A}$ randomly picks a location data $p_k = (x, y)$ from $DB_0$. According to the location encryption phase, $p_k$ is first extended to a $(n+10) \times (n+10)$-dimensional lower triangular random matrix $P_k$ with similar construction to $I_i$ with main diagonal elements $(0, 0, 1, x, \cdots, x^n, y, 0, 0, 0, 0, 0, 0, 0)$. Next, $\mathcal{A}$ encrypts $P_k$ to $C_k = A_k \times P_k$. Specifically, according to the algorithm of matrix multiplication, $C_k$ can be written as follows:

$$
C_k = \begin{bmatrix}
c_{1,1} & c_{1,2} & \cdots & c_{1,n+8} & 0 & 0 \\
c_{2,1} & c_{2,2} & \cdots & c_{2,n+8} & 0 & 0 \\
c_{3,1} & c_{3,2} & \cdots & c_{3,n+8} & 0 & 0 \\
\cdots & \cdots & \cdots & \cdots & \cdots & \cdots \\
c_{n+9,1} & c_{n+9,2} & \cdots & c_{n+9,n+8} & 0 & 0 \\
c_{n+10,1} & c_{n+10,2} & \cdots & c_{n+10,n+8} & 0 & 0
\end{bmatrix}_{(n+10)\times(n+10)} ,
$$

Let $\mathbf{A}_i$ be the $i$th row vector of $A_k$, $\mathbf{P}_j$ be the $j$th column vector of $P_k$, $a_{i,j}$ be the elements in the $i$th row and the $j$th column of $A_k$, and $p_{i,j}$ be the elements in the $i$th row and the $j$th column of $P_k$. Then, $c_{i,j}$ can be computed as follows:

$$
\begin{aligned}
c_{i,j} &= \mathbf{A}_i \circ \mathbf{P}_j \\
&= a_{i,1}p_{1,j} + a_{i,2}p_{2,j} + \cdots + a_{i,n+10}p_{n+10,j},
\end{aligned} \tag{4.15}
$$

where it can be known that

$$
\begin{cases}
p_{i,j} = *, \ 1 \geq j \geq i \geq n + 10, \ i \neq 2, \ j \neq n + 9, \\
p_{i,j} = x^{j-3}, \ 3 \geq j = i \geq n3, \\
p_{n+4,n+4} = y \\
p_{i,j} = 0, others.
\end{cases}
$$

Without loss of generality, we assume that $(x, y)$ can be any value that A chooses. Observe that the last column $\{c_{i,n+8}\}_{i=1}^{n+10}$ of $C_k$ is a non-zero vector, and we know that $c_{i,n+8} = a_{i,n+9}p_{n+9,n+8} + a_{i,n+10}p_{n+10,n+8}$. Since $a_{i,n+9}$ and $a_{i,n+10}$ are fixed numbers and $p_{n+9,n+8}$ and $p_{n+10,n+8}$ are random numbers chosen by $C$, it can be known that $c_{i,n+8}$ is a random number and changes with the change of $p_{n+9,n+8}$ and $p_{n+10,n+8}$. Next, we can know that $C_k$ is a random matrix, where each non-zero element contains at least two random numbers chosen by $C$.

In **Phase 1** and **Phase 2** of $Active_{\mathcal{A},GPTA}$, $\mathcal{A}$ can choose different $p_{i,j}$ each time and observe the corresponding ciphertext $C_{i,j}$. However, since $P_{i,j}$ is a random matrix generated by a different random number chosen by $C$ each time, the ciphertext obtained by $\mathcal{A}$ is completely random to $\mathcal{A}$. That is, given a ciphertext, $\mathcal{A}$ cannot determine which plaintext the ciphertext was encrypted by.

In summary, $\mathcal{A}$ can only determine the value of $b'$ by random guessing, and its advantage in guessing that $b' = b$ satisfies

$$
|Pr(Active_{A,GPTA}(\lambda) = 1) - \frac{1}{2}| \leq negl(\lambda). \tag{4.16}
$$

Thus, **Theorem 4.6** is proven.

### 4.4.2  Security Analysis of the Data Re-encryption Phase

In this section, we will analyze the security of GPTA in the data re-encryption phase. Without loss of generality, it is assumed that the cloud server $S_A$ may collude with some sensing users or data requesters and can run the game shown in Fig. 4.6. We have the following definitions and theorems.

**Definition 4.2** For any adversary $\mathcal{A}$ (the cloud server $Cloud$) with probabilistic polynomial time (P.P.T), it has a negligible advantage in guessing $b' = b$ as follows:

$$|Pr(Active_{A,GPTA}(\lambda) = 1) - \frac{1}{2}| \leq negl(\lambda), \tag{4.17}$$

and then GPTA can resist the active attack.

**Theorem 4.6** *GPTA can resist the active attack in the data re-encryption phase.*

**Proof** Suppose the sensing user encrypts the position $(x, y)$ as the ciphertext $C_i$. According to the proof of **Theorem 4.6**, $C_i$ is completely random to the adversary, so $S_A$ cannot determine the plaintext in the ciphertext. Since $S_A$ has a re-encryption key, $S_A$ can obtain $\tilde{C}_i = M_1 \times I_i \times (P_i + I_i') \times D_i \times M_A$ in the location re-encryption phase. If $M_1$ and $M_A$ are regarded as matrices similar to $A_i$, and $I_i$, $P_i$, $I_i'$, and $D_i$ are regarded as the lower triangular random matrices similar to $P_k$, according to the similar analysis of **Theorem 4.6**, it can be proved that GPTA can resist the active attack in the re-encryption phase.

It should be noted that since the system master key $M_1$ is included in the ciphertext $C_i$, if the cloud server obtains the master key $M_1$, it may obtain the user's private data. Next, we prove that even if the cloud server colludes with some sensing users or data requesters, it cannot recover the master key $M_1$ from the existing information.

Assuming that there are some user encryption keys $\{A_l\}_{l=1}^t$, $S_A$ computes $A' \times A_l$ to obtain a sequence of ciphertexts $F_l = M_1 \times I_1$. Let $f_{i,j}^{(l)}$, $m_{i,j}$, and $o_{i,j}^{(l)}$ be the $i$th row and $j$th column elements in $F_l$, $M_1$, and $I_l$, respectively. Without loss of generality, we choose the first element in the last non-zero column, i.e., $f_{1,n+8}^{(l)}$, and assume that $(x, y) = (0, 0)$. Then, it can be known that

$$\begin{cases} f_{1,n+8}^{(1)} = m_{i,n+9}o_{n+9,n+8}^{(1)} + m_{1,n+10}o_{n+10,n+8}^{(1)}, \\ f_{1,n+8}^{(2)} = m_{i,n+9}o_{n+9,n+8}^{(2)} + m_{1,n+10}o_{n+10,n+8}^{(2)}, \\ \cdots, \\ f_{1,n+8}^{(t)} = m_{i,n+9}o_{n+9,n+8}^{(t)} + m_{1,n+10}o_{n+10,n+8}^{(t)}. \end{cases}$$

In the above equations, there are $t$ equations. However, since there are $2t+2$ linearly independent unknowns (i.e., $m_{1,n+9}, m_{1,n+10}, \{o_{n+9,n+8}^{(l)}, o_{n+10,n+8}^{(l)}\}_{l=1}^t$), we cannot determine the unknowns $m_{1,n+9}$ and $m_{1,n+10}$. Thus, **Theorem 4.6** is proven.

### 4.4.3 Security Analysis of the Data Query Phase

Based on $\tilde{C}_i$, $\bar{T}_a$, and $\bar{T}_b$, the cloud $S_B$ can obtain $\Gamma_a = M_1 \times I_i \times (P_i + I_i') \times D_i \times M_A \times M_A^{-1} \times D_a \times (Q_a + S_q') \times S_q \times M_1^{-1}$ and $\Gamma_b = M_1 \times I_i \times (P_i + I_i') \times D_i \times M_A \times M_A^{-1} \times D_b \times (Q_b + S_q') \times S_q \times M_1^{-1}$. According to **Theorem 4.6**, since $M_1$ and $M_1^{-1}$ are unknown matrices, $I_i \times (P_i + I_i') \times D_i \times M_A \times M_A^{-1} \times D_a \times (Q_a + S_q') \times S_q$ and $I_i \times (P_i + I_i') \times D_i \times M_A \times M_A^{-1} \times D_b \times (Q_b + S_q') \times S_q$ are lower triangular random matrices. It can be known that $\Gamma_a$ and $\Gamma_b$ can also resist the active attack. That is, the cloud $S_B$ cannot obtain $l_i$, $l_a$, and $l_b$ from $\Gamma_a$ and $\Gamma_b$.

Next, observe the last results $\Delta_a = v_i v_a (\theta_a^*(x_i) - y_i) + r_{a,1}\gamma_{q,1}\gamma_{i,1} + r_{a,2}\gamma_{q,2}\gamma_{i,2}$ and $\Delta_b = v_i v_b (\theta_b^*(x_i) - y_i) + r_{b,1}\gamma_{q,1}\gamma_{i,1} + r_{b,2}\gamma_{q,2}\gamma_{i,2}$. Given a location data $l_i$, the cloud $S_B$ can choose two different queries $l_1$ and $l_b$ to obtain $\Delta_a > 0$ and $\Delta_b < 0$. In other words, the last results may reveal some information about $l_i$, $l_1$, and $l_2$. However, it is noted that the leaked results are what we need since we need $\Delta_a$ and $\Delta_b$ to determine whether a user is in the query range. $S_B$ may collude with some sensing users and data requesters. In this case, the values $SB$ knows are $_a^*(x_i)y_i$, $\Delta_a$ and $_b^*(x_i)y_i$, $\Delta_b$. Additionally, $\gamma_{i,1}$, $\gamma_{i,2}$, $\gamma_{q,1}$, $\gamma_{q,2}$ are unknowns, and $v_a$, $r_{a,1}$, $r_{a,2}$, $v_b$, $r_{b,1}$, $r_{b,2}$ are one-time random numbers related to $l_a$, $l_b$ generated by $S_A$. Thus, the last results can only reveal the positive and negative values of $\Delta_a$ and $\Delta_b$, and the cloud $S_B$ cannot obtain more information from it. In addition, when a user updates data or a new user joins, $S_B$ may speculate whether a user is within a certain range based on collusion trapdoors or previous historical trapdoors. GPTA can achieve forward privacy well because $S_A$ generates a different random number $(r_{i,1}, r_{i,2})$ for each newly added location data in the location re-encryption phase and $(r_{q,1}, r_{q,2})$ for each query trapdoor in the trapdoor re-encryption phase. Since the historical query trapdoors do not correlate with the number of newly added locations, $S_B$ cannot use the historical search trapdoor to determine whether the newly added location data is within the historical search range.

## 4.5 Performance Evaluation and Analysis

In this section, we conduct a theoretical analysis of GPTA and then implement GPTA to evaluate its performance. Specifically, we use a laptop with a 2.3 GHz Intel i5 processor and 16.0GB RAM as the cloud server and an Android phone with 4.0GB RAM as the data requester and sensing user. The latitude and longitude coordinates are randomly selected as the location of the sensing user, and the hyperbolic function that can cover several coordinates is randomly selected as the query range. To compare the performance, this section also implements the Paillier homomorphic encryption-based scheme ETA [9], the symmetric searchable encryption-based scheme PBRQ-L [14], and the random matrix multiplication technology-based scheme PPTA [25]. The experiments on the mobile phone side are implemented in Java, and the experiments on the laptop side are implemented in Python.

## 4.5.1   Query Accuracy

We first analyze the query accuracy of GPTA. GPTA uses the polynomial fitting technology to fit the query range. However, since the polynomial fitting technology is an approximate algorithm, it will inevitably introduce fitting errors. Therefore, this section analyzes the query accuracy by selecting different geometric figures (rectangles, circles, triangles, arbitrary range). We randomly pick points in [0, 1000] to simulate user coordinates. Table 4.2 presents the accuracy rates for different query ranges under different powers of the polynomial fitting function (i.e., $n$). It can be seen that with the increase of $n$, the fitting accuracy is getting higher and higher. When $n$ reaches 10, the fitting accuracy can reach 99% or even higher. In order to express the fitting accuracy more intuitively, Fig. 4.13, 4.14, 4.15, 4.16, 4.17, 4.18, 4.19, 4.20, 4.21, 4.22, 4.23, 4.24, 4.25, 4.26, 4.27, and 4.28 show the fitting graphs of different geometries under different $n$. It can be seen that GPTA can better fit different query ranges.

## 4.5.2   Theoretical Analysis

Table 4.3 shows the computational and communication overhead of GPTA-L and GPTA-F. For simplicity, we use $T_M$ to represent the computation time of the matrix multiplication operation, and the size of each matrix is denoted $|M|$. In the theoretical analysis, we only consider the computational cost of matrix operations and the communication overhead of matrix transmission. In order to encrypt a location in the GPTA-L and GPTA-F, a sensing user needs to perform a matrix multiplication operation in the location encryption phase. The cloud server needs to perform three matrix multiplication operations in the location re-encryption phase. Thus, the computational costs on the sensing user and cloud side are $T_M$ and $3T_M$, respectively. In the trapdoor generation phase, GPTA-L needs to perform two matrix multiplication operations, while GPTA-F needs to perform two more matrix multiplication operations since GPTA-F needs to encrypt the circle that

**Table 4.2** Query accuracy of GPTA at different degrees

|                 | $n = 3$ | $n = 5$ | $n = 7$ | $n = 10$ |
|-----------------|---------|---------|---------|----------|
| Triangle        | 93.72%  | 98.52%  | 98.00%  | 99.15%   |
| Circle          | 98.11%  | 99.29%  | 99.64%  | 99.80%   |
| Rectangle       | 93.87%  | 97.03%  | 98.21%  | 99.12%   |
| Arbitrary range | 83.09%  | 95.17%  | 99.36%  | 99.99%   |

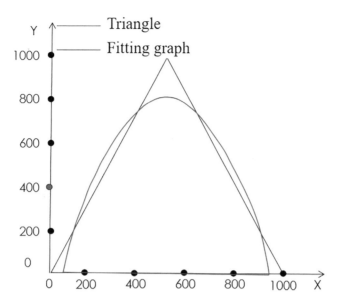

**Fig. 4.13** Triangle fitting with $n = 3$

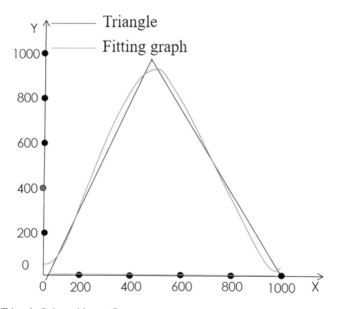

**Fig. 4.14** Triangle fitting with $n = 5$

covers the query range. Correspondingly, GPTA-L and GPTA-F need to perform 8 and 15 matrix multiplications, respectively, in the trapdoor re-encryption phase. In the query phase, assuming the number of encrypted ciphertexts is $m$, the computational cost of GPTA-L is $2mT_M$. For GPTA-F, since GPTA-F utilizes a

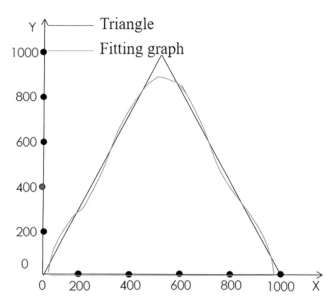

**Fig. 4.15** Triangle fitting with $n = 7$

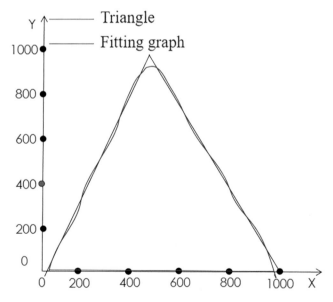

**Fig. 4.16** Triangle fitting with $n = 10$

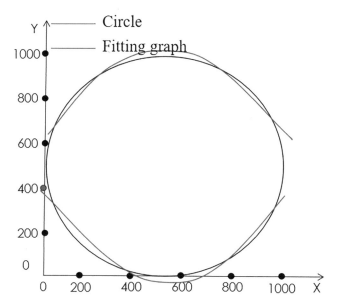

**Fig. 4.17**  Circle fitting with $n = 3$

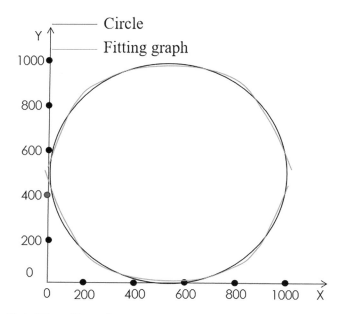

**Fig. 4.18**  Circle fitting with $n = 5$

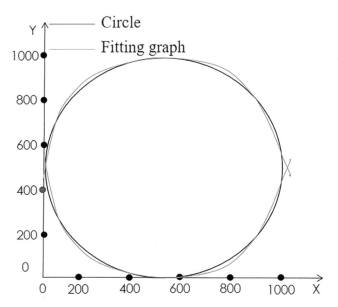

**Fig. 4.19**  Circle fitting with $n = 7$

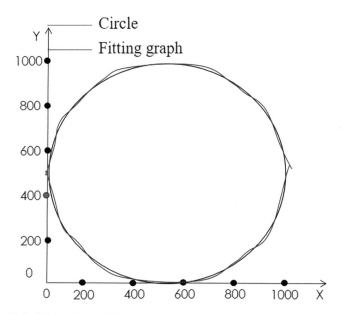

**Fig. 4.20**  Circle fitting with $n = 10$

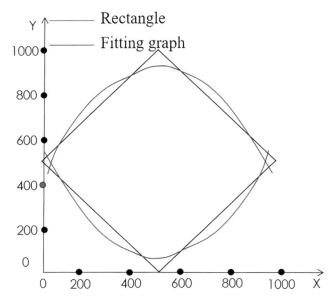

**Fig. 4.21** Rectangular fitting with $n = 3$

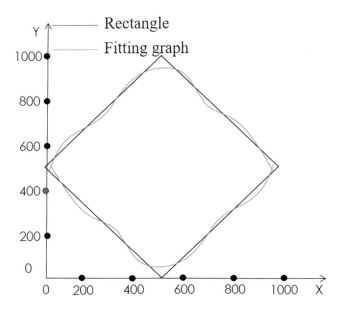

**Fig. 4.22** Rectangular fitting with $n = 5$

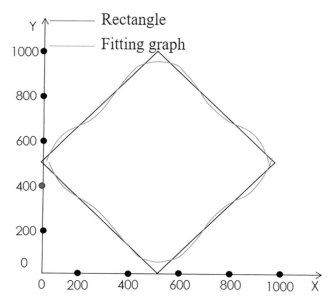

**Fig. 4.23**  Rectangular fitting with $n = 7$

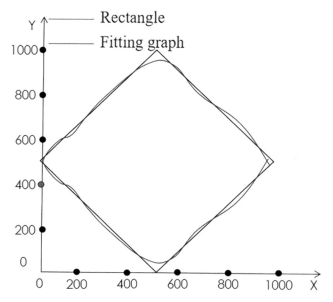

**Fig. 4.24**  Rectangular fitting with $n = 10$

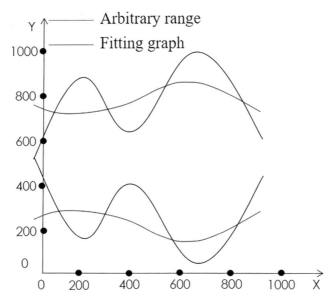

**Fig. 4.25**  Arbitrary range fitting with $n = 3$

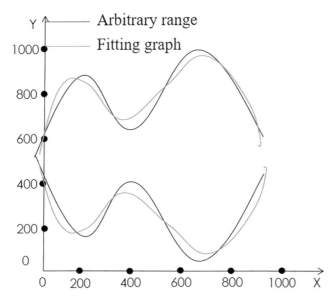

**Fig. 4.26**  Arbitrary range fitting with $n = 5$

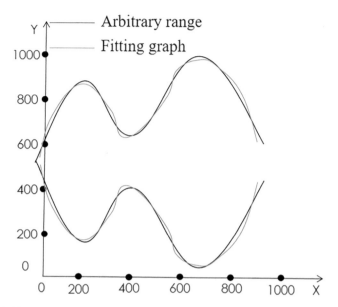

**Fig. 4.27**  Arbitrary range fitting with $n = 7$

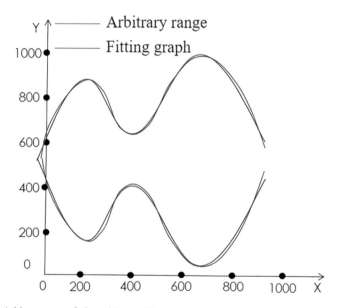

**Fig. 4.28**  Arbitrary range fitting with $n = 10$

**Table 4.3** Theoretical analysis of GPTA-L and GPTA-F in computational cost and communication overhead

| Scheme | Phase | Entity | Computation | Communication |
|---|---|---|---|---|
| GPTA-L | Location encryption | Sensing user | $T_M$ | $|M|$ |
| | Location re-encryption | Platform | $3T_M$ | $|M|$ |
| | Trapdoor generation | Data requester | $2T_M$ | $2|M|$ |
| | Trapdoor re-encryption | Platform | $8T_M$ | $2|M|$ |
| | Query | Platform | $2mT_M$ | – |
| GPTA-F | Task encryption | Sensing user | $T_M$ | $|M|$ |
| | Task re-encryption | Platform | $3T_M$ | $|M|$ |
| | Trapdoor generation | Data requester | $4T_M$ | $4|M|$ |
| | Trapdoor re-encryption | Platform | $15T_M$ | $4|M|$ |
| | Query | Platform | $(2(m-Kz-ez)+K)T_M$ | – |

**Table 4.4** Theoretical analysis in computational cost and communication overhead between GPTA and other works

| Scheme | Phase | Entity | Computation | Communication |
|---|---|---|---|---|
| GPTA-L | Location encryption | Sensing user | $T_M$ | $|M|$ |
| | Trapdoor generation | Data requester | $2T_M$ | $2|M|$ |
| | Query | Platform | $2mT_M$ | – |
| GPTA-F | Task encryption | Sensing user | $T_M$ | $|M|$ |
| | Trapdoor generation | Data requester | $4T_M$ | $4|M|$ |
| | Query | Platform | $(2(m-Kz-ez)+K)T_M$ | – |
| PBRQ-L | Task encryption | Sensing user | $wT_{PRF}$ | $w|PRF|$ |
| | Trapdoor generation | Data requester | $wT_{PRF}+T_{Enc}$ | $|PRF|$ |
| | Query | Platform | $m(wT_{XOR}+T_{Dec})$ | – |
| ETA | Task encryption | Sensing user | $2P_{Enc}$ | $2|Paillier|$ |
| | Trapdoor generation | Data requester | $2P_{Enc}$ | $2|Paillier|$ |
| | Query | Platform | $m(5P_{Enc}+P_{Dec})$ | $(10m+4)|Paillier|$ |
| PPTA | Task encryption | Sensing user | $\widetilde{T}_M$ | $|\widetilde{M}|$ |
| | Trapdoor generation | Data requester | $k\widetilde{T}_M$ | $k|\widetilde{M}|$ |
| | Query | Platform | $km\widetilde{T}_M$ | – |

novel index structure to speed up the query time, GPTA-F will perform fewer matching operations if there are more parent nodes. Assuming that there are $K$ parent nodes in total, where there are $e$ parent nodes that intersect with the new query range, and each parent node is assumed to have $z$ nodes, then the cloud server needs to perform $(2(m-Kz-ez)+K)T_M$ matrix multiplication operations.

The comparison between GPTA and the existing schemes [9, 14, 25] is summarized in Table 4.4. Among them, scheme [14] is implemented based on the symmetric encryption algorithm SHVE [30], which uses a vector that contains $w$ bits to represent the location of a sensing user. In scheme [14], we use $T_{PRF}$ to represent the time to calculate a pseudo-random function, $T_{XOR}$ to represent the time to perform an XOR operation, $T_{Enc}$ to represent the time to perform symmetric algorithm encryption, $T_{Dec}$ to represent the time to perform symmetric algorithm decryption, and $|PRF|$ to represent the size of the output of a pseudo-random function. Scheme [9] is implemented based on the Paillier encryption algorithm, which uses latitude and longitude to represent the location of the sensing user. In this scheme [9], we use $P_{Enc}$ to represent the time to perform a Paillier encryption operation, $P_{Dec}$ to represent the time to perform a Paillier decryption operation, and $|Paillier|$ to represent the size of a Paillier ciphertext. Scheme [25] is implemented based on random matrix multiplication, which uses the elements in the matrix to represent a geometric area. Suppose that $W$ is the number of all geometric regions and the dimension of the matrix is equal to $\sqrt{W}$. In this scheme [25], we use $\widetilde{TM}$ to denote the time to perform a matrix multiplication operation, and $|\widetilde{M}|$ to denote the size of a matrix. The number of geometric regions requested by the data requester is $k$. It should be pointed out that the schemes [9, 14] are implemented based on symmetric encryption or Paillier homomorphic encryption, while GPTA is implemented based on random matrix multiplication, which can save more computational cost. Although scheme [25] is also implemented based on random matrix multiplication, this scheme takes more time to perform data encryption and query operations than GPTA. Specifically, in scheme [25], since the dimension of the matrix in this scheme increases with the increase of the number of geometrical areas, the computational cost will be very high when the geometrical range is large.

### 4.5.3   Performance Evaluation

**Experimental Environment and Implementation**   In scheme [14], the quadtree and Gray code are used to represent a user's location, and SHVE [30] is used to preserve location privacy. In order to achieve the arbitrary query range, the size of the Gray code should be set large enough. We set the length of the Gray code to 100 and the size of the symmetric encryption key to 128 bits. In scheme [9], the key of Paillier's homomorphic encryption algorithm is set to 512 bits. In scheme [25], the total number of geometric regions, i.e., $W$, is set to 10000, and the number of geometric regions required by the data requester is set to 1. The highest power of the curve equation generated by the polynomial fitting technology in GPTA is set to 10. The value range of each element in the matrix is $[1, 2^{32}]$. The total number of encrypted locations is $1 \times 10^5$ to $1 \times 10^6$. The location encryption and trapdoor generation phases are implemented on the mobile phone, and the other phases are implemented on the laptop.

**Experimental Results**   Figures 4.29 and 4.30 show the computational cost of all schemes in the phase of location encryption and trapdoor generation. It can be seen that the GPTA scheme is more efficient than the other three schemes. This is because (1) GPTA is implemented based on random matrix multiplication, while schemes

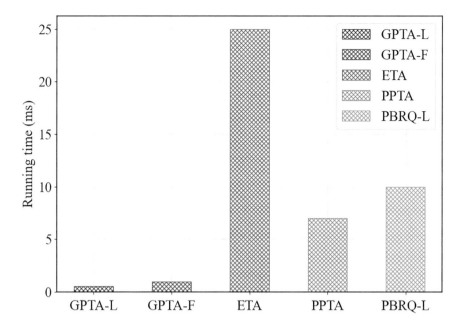

**Fig. 4.29**  Computational cost in the location encryption phase

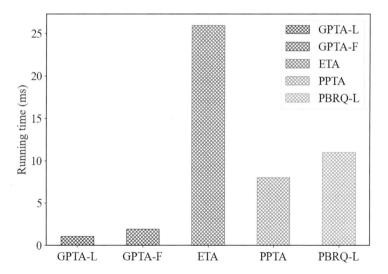

**Fig. 4.30**  Computational cost in the phase trapdoor generation phase

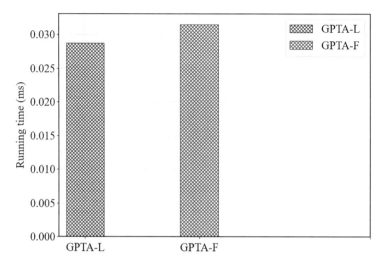

**Fig. 4.31** Computational cost in the location re-encryption phase

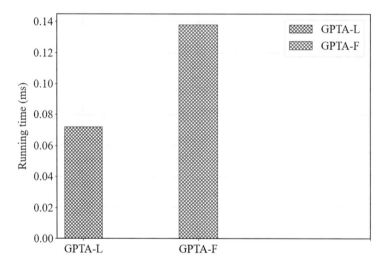

**Fig. 4.32** Computational cost in the phase trapdoor re-encryption phase

[9, 14] are implemented based on traditional cryptographic encryption algorithms, and (2) due to the use of the polynomial fitting technology, the dimensions of the matrices in GPTA remain unchanged, while the size of the data that needs to be encrypted in schemes [14, 25] increases with the number of geometric regions.

Figures 4.31 and 4.32 show the computational costs of the location re-encryption and trapdoor re-encryption phases. From Figs. 4.31 and 4.32, it can be seen that the computational cost of GPTA-F is larger than that of GPTA-L because GPTA-F needs to perform additional re-encryption operations on the newly generated circle.

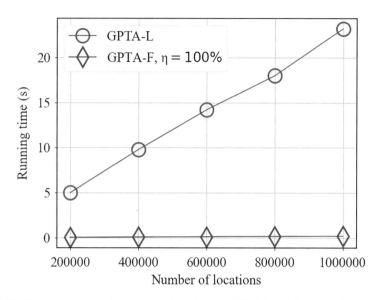

**Fig. 4.33** Computational cost in the query phase between GPTA-L and GPTA-F with $\eta = 100\%$

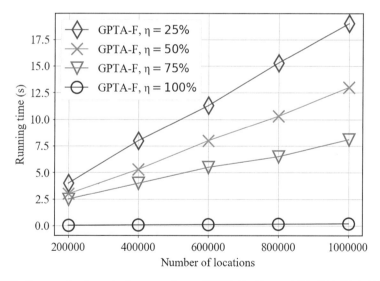

**Fig. 4.34** Computational cost in the query phase of GPTA-F with $\eta = 25\%, 50\%, 75\%, 100\%$

From the experimental results, it can be seen that the re-encryption operations for both the location and the trapdoor can be efficiently completed in 0.14 ms by both schemes.

Figures 4.33 and 4.34 illustrate the computational costs of the query phase, where the number $m$ of encrypted locations is set from $2 \times 10^5$ to $1 \times 10^6$. For a geometric range query, the query time of GPTA-L increases with the number of encrypted locations. GPTA-F designs a new index structure to improve query time. To more clearly illustrate the impact of the index structure, we use $\eta = \frac{zK}{m}$ to denote the number of child nodes in the parent node. Specifically, we assume $z = 100$, $e = 1$, $\eta \in [25\%, 100\%]$. As shown in Fig. 4.33, when $\eta = 100\%$, where all sensing users are located in the parent node, the time cost of GPTA-F can save about 98% compared to GPTA-L. When choosing different $\eta$, as shown in Fig. 4.34, the query time of GPTA-F is inversely proportional to the value of $\eta$. That is, as the number of queries increases, more and more nodes will be added to the parent node so that the query time becomes less and less. Therefore, GPTA-F can be proved to be a scheme with nonlinear query efficiency.

The computational costs of GPTA and schemes [9, 14, 25] in the user query phase are given in Table 4.5. It can be seen that GPTA is more efficient. In particular, GPTA-L with linear query efficiency is less computationally expensive than the other three schemes. The results in Table 4.5 verify that GPTA has high computational efficiency.

In addition, the communication overhead of GPTA and schemes [9, 14, 25] at different phases are given in Table 4.6, where $h$ represents hours and $s$ represents seconds. It can be seen that GPTA consumes more communication overhead than ETA and PBRQ-L but saves more communication overhead than PPTA. This is because GPTA uses the polynomial fitting technology to fit the range, so the dimensions of the matrix are not affected by the geometric range. Because ETA finds the closest sensing user for the data requester, it only needs to encrypt the user's location. Since PBRQ-L uses a pseudo-random function to preserve the user's

**Table 4.5** Computational cost comparison between GPTA and other works

| Scheme | $1 \times 10^5$ | $2 \times 10^5$ | $3 \times 10^5$ | $4 \times 10^5$ | $5 \times 10^5$ |
|---|---|---|---|---|---|
| ETA | 0.79 h | 1.69 h | 2.63 h | 3.42 h | 4.11 h |
| PPTA | 49.78 s | 99.39 s | 148.45 s | 202.04 s | 286.35 s |
| PBRQ-L | 3.18 s | 6.36 s | 9.89 s | 12.85 s | 15.73 s |
| GPTA-L | 2.92 s | 5.56 s | 7.84 s | 10.36 s | 13.56 s |
| GPTA-F (25%) | 1.97 s | 3.75 s | 5.59 s | 7.41 s | 9.22 s |
| GPTA-F (50%) | 1.43 s | 2.68 s | 3.86 s | 5.34 s | 6.86 s |
| GPTA-F (75%) | 1.23 s | 2.35 s | 2.86 s | 3.61 s | 4.60 s |
| GPTA-F (100%) | 0.02 s | 0.04 s | 0.06 s | 0.09 s | 0.12 s |

**Table 4.6** Communication overhead comparison between GPTA and other works

| Scheme | Location encryption $u_i - S_A$ | Trapdoor generation $u_q - S_A$ |
|---|---|---|
| ETA | 0.5 KB | 0.5 KB |
| PPTA | 39.8 KB | 39.8 KB |
| PBRQ-L | 1.56 KB | 0.02 KB |
| GPTA-L | 1.59 KB | 3.18 KB |
| GPTA-F | 1.59 KB | 6.36 KB |

location and uses an exclusive XOR operation to generate trapdoors, it will save a large amount of communication overhead.

## 4.6  Summary

Geometric range-based task allocation can help data requesters find sensing users who meet their geometric range requirements. However, users' geometric location information may reveal user privacy. In this chapter, we study the security and efficiency of geometric range-based privacy-preserving task allocation in mobile crowdsensing scenarios and propose two privacy-preserving task allocation schemes. Specifically, we firstly design a basic geometric range-based privacy-preserving task allocation scheme (GPTA-L) that supports arbitrary geometric range based on random matrix multiplication technology and polynomial fitting technology. Then, by studying the historical query behaviors of data requesters, we propose an improved scheme GPTA-F with nonlinear query efficiency. In particular, the schemes designed in this chapter are constructed based on the non-collusive server model, which can resist the collusion attack launched by a single server and sensing users or data requesters. Security analysis proves that GPTA can effectively preserve the location privacy of sensing users and the query privacy of data requesters. Extensive experiments demonstrate the high computational efficiency of GPTA in data encryption and data query. On the basis of this chapter, considering the malicious behavior of cloud servers, it is a meaningful research direction to study a verifiable geometric range-based privacy-preserving task allocation scheme based on the non-collusive server model.

## References

1. Gil, D.S., P.M. d'Orey, Aguiar, A.: On the challenges of mobile crowdsensing for traffic estimation. In: Eskicioglu, M.R. (ed.) Proceedings of the 15th ACM Conference on Embedded Network Sensor Systems, SenSys 2017, Delft, November 06–08, 2017, pp. 52:1–52:2. ACM, New York (2017)
2. Yan, H., Hua, Q., Zhang, D., Wan, J., Rho, S., Song, H.: Cloud-assisted mobile crowd sensing for traffic congestion control. Mob. Netw. Appl. **22**(6), 1212–1218 (2017)
3. Fiandrino, C., Capponi, A., Cacciatore, G., Kliazovich, D., Sorger, U., Bouvry, P., Kantarci, B., Granelli, F., Giordano, S.: Crowdsensim: a simulation platform for mobile crowdsensing in realistic urban environments. IEEE Access **5**, 3490–3503 (2017)
4. Koukoutsidis, I.Z.: Estimating spatial averages of environmental parameters based on mobile crowdsensing. ACM Trans. Sens. Netw. **14**(1), 2:1–2:26 (2018)
5. Huang, P., Zhang, X., Guo, L., Li, M.: Incentivizing crowdsensing-based noise monitoring with differentially-private locations. IEEE Trans. Mob. Comput. **20**(2), 519–532 (2021)
6. Liu, Y., Ma, X., Shu, L., Yang, Q., Zhang, Y., Huo, Z., Zhou, Z.: Internet of things for noise mapping in smart cities: state of the art and future directions. IEEE Netw. **34**(4), 112–118 (2020)

7. Wang, Z., Hu, J., Lv, R., Wei, J., Wang, Q., Yang, D., Qi, H.: Personalized privacy-preserving task allocation for mobile crowdsensing. IEEE Trans. Mob. Comput. **18**(6), 1330–1341 (2019)
8. Zhu, H., Lu, R., Huang, C., Chen, L., Li, H.: An efficient privacy-preserving location-based services query scheme in outsourced cloud. IEEE Trans. Veh. Technol. **65**(9), 7729–7739 (2016)
9. Liu, A., Li, Z., Liu, G., Zheng, K., Zhang, M., Li, Q., Zhang, X.: Privacy-preserving task assignment in spatial crowdsourcing. J. Comput. Sci. Technol. **32**(5), 905–918 (2017)
10. Huang, C., Lu, R., Zhu, H., Shao, J., Alamer, A., Lin, X.: EPPD: efficient and privacy-preserving proximity testing with differential privacy techniques. In: 2016 IEEE International Conference on Communications, ICC 2016, Kuala Lumpur, May 22–27, 2016, pp. 1–6. IEEE, Piscataway (2016)
11. Xu, G., Li, H., Dai, Y., Yang, K., Lin, X.: Enabling efficient and geometric range query with access control over encrypted spatial data. IEEE Trans. Inf. Forensics Secur. **14**(4), 870–885 (2019)
12. Wang, B., Li, M., Wang, H.: Geometric range search on encrypted spatial data. IEEE Trans. Inf. Forensics Secur. **11**(4), 704–719 (2016)
13. Wang, B., Li, M., Xiong, L.: Fastgeo: efficient geometric range queries on encrypted spatial data. IEEE Trans. Dependable Secure Comput. **16**(2), 245–258 (2019)
14. Wang, X., Ma, J., Liu, X., Deng, R.H., Miao, Y., Zhu, D., Ma, Z.: Search me in the dark: privacy-preserving boolean range query over encrypted spatial data. In: 39th IEEE Conference on Computer Communications, INFOCOM 2020, Toronto, July 6–9, 2020, pp. 2253–2262. IEEE, Piscataway (2020)
15. Wang, B., Li, M., Wang, H., Li, H.: Circular range search on encrypted spatial data. In: 2015 IEEE Conference on Communications and Network Security, CNS 2015, Florence, September 28–30, 2015, pp. 182–190. IEEE, Piscataway (2015)
16. Zhang, X., Lu, R., Shao, J., Zhu, H., Ghorbani, A.A.: Secure and efficient probabilistic skyline computation for worker selection in MCS. IEEE Internet Things J. **7**(12), 11524–11535 (2020)
17. Cui, N., Li, J., Yang, X., Wang, B., Reynolds, M., Xiang, Y.: When geo-text meets security: privacy-preserving boolean spatial keyword queries. In: 35th IEEE International Conference on Data Engineering, ICDE 2019, Macao, April 8–11, 2019, pp. 1046–1057. IEEE, Piscataway (2019)
18. Zheng, Y., Lu, R., Guan, Y., Shao, J., Zhu, H.: Achieving efficient and privacy-preserving exact set similarity search over encrypted data. IEEE Trans. Dependable Secur. Comput. **19**(2), 1090–1103 (2022)
19. Zhai, D., Sun, Y., Liu, A., Li, Z., Liu, G., Zhao, L., Zheng, K.: Towards secure and truthful task assignment in spatial crowdsourcing. World Wide Web **22**(5), 2017–2040 (2019)
20. Wang, L., Yang, D., Han, X., Wang, T., Zhang, D., Ma, X.: Location privacy-preserving task allocation for mobile crowdsensing with differential geo-obfuscation. In: Barrett, R., Cummings, R., Agichtein, E., Gabrilovich, E. (eds.) Proceedings of the 26th International Conference on World Wide Web, WWW 2017, Perth, April 3–7, 2017, pp. 627–636. ACM, New York (2017)
21. Tang, W., Zhang, K., Ren, J., Zhang, Y., (Sherman) Shen, X.: Privacy-preserving task recommendation with win-win incentives for mobile crowdsourcing. Inf. Sci. **527**, 477–492 (2020)
22. Shu, J., Liu, X., Jia, X., Yang, K., Deng, R.H.: Anonymous privacy-preserving task matching in crowdsourcing. IEEE Internet Things J. **5**(4), 3068–3078 (2018)
23. Hao, J., Huang, C., Chen, G., Xian, M., Shen, X.S.: Privacy-preserving interest-ability based task allocation in crowdsourcing. In: 2019 IEEE International Conference on Communications, ICC 2019, Shanghai, May 20–24, 2019, pp. 1–6. IEEE, Piscataway (2019)
24. Zhang, J., Zhang, Q., Ji, S.: A fog-assisted privacy-preserving task allocation in crowdsourcing. IEEE Internet Things J. **7**(9), 8331–8342 (2020)
25. Ni, J., Zhang, K., Xia, Q., Lin, X., Shen, X.S.: Enabling strong privacy preservation and accurate task allocation for mobile crowdsensing. IEEE Trans. Mob. Comput. **19**(6), 1317–1331 (2020)

26. Cao, N., Wang, C., Li, M., Ren, K., Lou, W.: Privacy-preserving multi-keyword ranked search over encrypted cloud data. IEEE Trans. Parallel Distrib. Syst. **25**(1), 222–233 (2014)
27. Shen, E., Shi, E., Waters, B.: Predicate privacy in encryption systems. In: Reingold, O. (ed.) Proceedings of the Theory of Cryptography, 6th Theory of Cryptography Conference, TCC 2009, San Francisco, March 15–17, 2009, vol. 5444.Lecture Notes in Computer Science, pp. 457–473. Springer, Berlin (2009)
28. Song, D.X., Wagner, D.A., Perrig, A.: Practical techniques for searches on encrypted data. In: 2000 IEEE Symposium on Security and Privacy, Berkeley, May 14–17, 2000, pp. 44–55. IEEE Computer Society, Washington (2000)
29. Miao, Y., Deng, R., Choo, K.-K.R., Liu, X., Li, H.: Threshold multi-keyword search for cloud-based group data sharing. IEEE Trans. Cloud Comput. 1–1 (2020)
30. Lai, S., Patranabis, S., Sakzad, A., Liu, J.K., Mukhopadhyay, D., Steinfeld, R., Sun, S., Liu, D., Zuo, C.: Result pattern hiding searchable encryption for conjunctive queries. InL Proceedings of the 2018 ACM SIGSAC Conference on Computer and Communications Security, CCS 2018, Toronto, October 15–19, 2018, pp. 745–762. ACM, New York (2018)
31. Xu, G., Li, H., Dai, Y., Yang, K., Lin, X.: Enabling efficient and geometric range query with access control over encrypted spatial data. IEEE Trans. Inf. Forensics Secur. **14**(4), 870–885 (2018)
32. Strobach, P.: Solving cubics by polynomial fitting. J. Comput. Appl. Math. **235**(9), 3033–3052 (2011)
33. Miao, C., Su, L., Jiang, W., Li, Y., Tian, M.: A lightweight privacy-preserving truth discovery framework for mobile crowd sensing systems. In: 2017 IEEE Conference on Computer Communications, INFOCOM 2017, Atlanta, May 1–4, 2017, pp. 1–9. IEEE, Piscataway (2017)
34. Liu, X., Qin, B., Deng, R.H., Lu, R., Ma, J.: A privacy-preserving outsourced functional computation framework across large-scale multiple encrypted domains. IEEE Trans. Comput. **65**(12), 3567–3579 (2016)
35. Zhang, C., Zhu, L., Xu, C., Liu, X., Sharif, K.: Reliable and privacy-preserving truth discovery for mobile crowdsensing systems. IEEE Trans. Dependable Secur. Comput. **18**(3), 1245–1260 (2021)
36. Xue, K., Li, S., Hong, J., Xue, Y., Yu, N., Hong, P.: Two-cloud secure database for numeric-related SQL range queries with privacy preserving. IEEE Trans. Inf. Forensics Secur. **12**(7), 1596–1608 (2017)
37. Xu, J., Xue, K., Yang, Q., Hong, P.: PSAP: pseudonym-based secure authentication protocol for NFC applications. IEEE Trans. Consumer Electron. **64**(1), 83–91 (2018)
38. Yang, Q., Xue, K., Xu, J., Wang, J., Li, F., Yu, N.: AnFRA: anonymous and fast roaming authentication for space information network. IEEE Trans. Inf. Forensics Secur. **14**(2):486–497 (2019)

# Part III
# Privacy-Preserving Truth Discovery

# Chapter 5
# Privacy-Preserving Truth Discovery with Truth Transparency

**Abstract** This chapter will introduce how to identify the data submitted by sensing users in the data processing phase and calculate the truth with privacy-preserving. For the same task, the data submitted by sensing users often varies widely, and user concerns about the privacy of their data may affect the use of truth discovery. Existing truth discovery schemes for privacy-preserving either have large computational and communication costs or cannot provide comprehensive protection for user data and weight. Therefore, based on a homomorphic encryption algorithm and two-server model, this chapter proposes a truth discovery scheme RPTD for privacy protection. Specifically, the chapter first proposes a privacy-preserving truth discovery scheme RPTD-I with user participation in the truth discovery iteration process, which can effectively preserve the privacy of user's data and weights and achieve efficient and secure truth updates and weight updates. Subsequently, this chapter proposes a privacy-preserving truth discovery scheme RPTD-II without user participation in the truth discovery iteration process, which transfers the user's computation in the truth discovery iteration process to a cloud server, greatly reducing the user's computational and communication overhead.

## 5.1 Introduction

This section describes the overview, related works, and preliminary of this chapter.

### 5.1.1 Overview

In recent years, with the rapid development of mobile terminal devices (such as smartphones, smart glasses, smart watches, etc.), mobile crowdsensing (MCS) has become an effective mode of data collection and analysis [1, 2]. By assigning data sensing tasks (e.g., health data collection, location information collection, etc.) to a large number of sensing users, crowdsensing can serve large-scale and complex social sensing activities such as health sensing [3, 4], urban sensing [5], and environmental monitoring [6]. However, the data collected by different sensing devices may vary greatly due to factors such as sensor quality, background noise, device movement, and user subjective awareness. Therefore, an important part of data processing is how to solve for the true results (called true values) from the user's sensory data; the process is called "data truth seeking."

An intuitive way to find the truth of the data is to assign different weights to different users and then solve for the true value by weighting the average. For example, if a user has a better quality perception device or a deeper knowledge of a domain, the user will be given a higher weight and more likely the user's data will be considered as the true value. However, in the actual crowdsensing scenario, the user's weight is usually unknown, and the mobile crowdsensing platform cannot infer the user's weight information from the existing information. To solve this problem, truth discovery, as an effective technique that can compute true values in sensory data, has attracted the attention of more and more scholars in recent years. Truth discovery is an iterative algorithm containing true value update and weight update, whose basic principles are (1) weight update—if a user provides data closer to the true value, that user will be given a higher weight and (2) true value update—if a user has a higher weight, that user's data will have a higher weight in the calculation of the true value.

Existing schemes on truth discovery [7–13] mainly focus on the efficiency and accuracy of data truth finding, but rarely consider the security and privacy issues that exist in it. In crowdsensing applications, the sensory data submitted by users usually contain private information related to them, such as location, behavioral habits, health status, etc. [14]. Mobile crowdsensing platforms or other entities may violate users' privacy based on their data or misuse users' data to gain more financial benefits. For example, providing rating data of a restaurant helps other users to know the overall situation of the restaurant, but the mobile crowdsensing platform may infer the user's hobbies, financial status, etc. based on the user's rating, and then target the user with advertising push. In addition to sensing data, the user weights solved by applying truth discovery may also reveal the user's privacy. For example, if a user participates in a difficult question-answering task issued by a mobile crowdsensing platform (e.g., Zhihu, Baidu Knows, etc.) and if the user has a high weight, then the user is more knowledgeable about the relevant question, and the mobile crowdsensing platform may infer the user's identity, occupation, hobbies, gender, faith, education, and other sensitive information and then target advertising to the user or sell the user's private information to other platforms to seek more economic benefits. Therefore, the user's data and weight privacy should be strictly protected in the process of data truth seeking.

Intuitively, we can use traditional encryption algorithms to protect users' privacy and design relevant protocols to achieve privacy-preserving truth discovery. However, considering that truth discovery is an iterative update algorithm and mobile devices are characterized by high mobility and limited computational and communication performance, the efficiency issue has become an important challenge that limits the application of privacy-preserving truth discovery schemes. Currently, scholars have devised a series of privacy-preserving truth discovery schemes

[15–19]; however, most of these schemes have efficiency or privacy concerns. For example, Miao et al. [15] proposed a privacy-preserving truth discovery scheme in a cloud computing environment using the threshold Paillier cryptosystem [20]. Although this scheme can effectively protect the user's identity and weight privacy, each user needs to perform time-consuming encryption operations and upload a large number of ciphertexts to support the iterative update of weights and truth values. To achieve more efficient privacy-preserving truth discovery, Xu et al. [17] designed a privacy-preserving truth discovery scheme using a lightweight privacy-preserving aggregation protocol. This scheme has higher computational and communication efficiency than scheme [15] due to the use of random numbers to protect the user's privacy. However, since the scheme requires the user to upload the ciphertext on time at each iteration, it may be limited in practical applications. Miao et al. [16] proposed a privacy-preserving truth discovery scheme based on a non-collusive dual server. This scheme can transfer most of the user's computation process to the cloud server and greatly reduce the user's computational and communication overhead. However, this scheme does not consider the privacy of weights. Zheng et al. [19] implemented privacy-preserving true value discovery in ciphertext form using garbled circuit (GC). However, since the cloud server needs to generate garbled circuits, this scheme suffers a large impact on the computational and communication efficiency when the number of users or tasks is large.

To address the above challenges, this chapter proposes a reliable and privacy-preserving truth discovery (RPTD) scheme based on a non-collusive dual server. Specifically, this chapter first proposes the privacy-preserving truth discovery scheme RPTD-I for the scenario where the user's location is relatively fixed and can participate in the iterative truth discovery process. Considering that some users may be in a frequently moving environment, such as connected cars, this chapter proposes a user-free truth discovery iterative scheme RPTD-II based on RPTD-I. The contributions of this chapter are summarized as follows:

- In this chapter, we first propose scheme RPTD-I to solve the privacy and efficiency problems in mobile group-wise sensing truth discovery. In this scheme, the sensing user participates in the iterative update process of true value discovery, and the privacy-preserving true value discovery is achieved by homomorphic encryption algorithm PCDD and superlinear sequence.
- For the scenario where users move frequently, this chapter proposes scheme RPTD-II based on RPTD-I. In this scheme, users only need to submit data once, and all the computations of users in the iterative process of true value discovery are transferred to the cloud server.
- The security and privacy analysis demonstrates that RPTD can effectively protect the user's data and weight privacy. Experiments based on real and simulated mobile swarm intelligence awareness scenarios validate that RPTD has high computational and communication efficiency.

Note that this chapter and most existing privacy-preserving truth discovery schemes [15–19] consider privacy-preserving truth discovery with truth disclosure, i.e., RPTD allows the last computed true value to be disclosed to the mobile crowdsensing platform. This scheme is applicable to the scenario where the mobile crowdsensing platform is the data requester. In practical MCS applications, mobile crowdsensing platforms (e.g., Meituan, Zhihu, WAZE, etc.) may collect various perception data by employing a large number of perception users to obtain truth values for different perception tasks, and then sell the true values to different organizations or individuals to gain benefits. In the next chapter, we will consider the scenario where data requesters are involved in data truth seeking and design a privacy-preserving truth discovery scheme for truth hiding.

The rest of this chapter is organized as follows: Sect. 5.1.2 reviews the related works on privacy-preserving truth discovery; Sect. 5.1.3 provides a brief description of preliminaries; Sect. 5.2 describes the architecture overview of RPTD from system model, security model, and design goals; Sect. 5.3 presents the detailed design flow of RPTD; Sect. 5.4 gives the security analysis of RPTD-I and RPTD-II; Sect. 5.5 presents the performance evaluation and analysis; and Sect. 5.6 summarizes the work of this chapter.

## 5.1.2  Related Works

As a data truth finding technique, truth discovery has attracted the attention of more and more researchers in recent years and has been widely used in different scenarios or applications [11, 12, 21]. For example, Li et al. [21] proposed a generic truth discovery algorithm to compute true values from unreal data for heterogeneous data types, and Meng et al. [11] designed a corresponding truth discovery algorithm for correlation between entities to improve the accuracy and efficiency of data truth finding. All the above truth discovery schemes do not consider the privacy of user data and weights. To achieve privacy-preserving truth discovery, Miao et al. [15] used a threshold Paillier encryption algorithm to protect the user's sensory data and weight information. However, due to time-consuming cryptographic operations, this scheme imposes significant computational and communication overhead for both the user and the mobile crowdsensing platform. Xu et al. [17] achieved efficient true value discovery in ciphertext form using data aggregation protocols [22] and superlinear sequences [23, 24]. Specifically, the trusted center assigns a random number to each sensing user to perturb the sensing data, and the sum of the random numbers is sent to the mobile crowdsensing platform. The mobile crowdsensing platform aggregates the perturbed data and then subtracts the sum of the random numbers to obtain the true aggregation result. However, since the sum of random numbers is fixed, this scheme gives wrong results if some users fail to upload data on time. Miao et al. [16] proposed a privacy-preserving truth discovery scheme $L^2 - PPTD$ based on two disjoint cloud servers, which does not require users to participate in the truth discovery iteration process. Although this scheme can greatly reduce the computational and communication overhead of users, the privacy of users' weights is not well protected. Zheng et al. [18] also proposed targeted privacy-preserving truth discovery schemes for single server and

**Table 5.1** Comparison of the functional and privacy attributes of RPTD and existing solutions

| Scheme | Whether users need to participate | Sensory data privacy | Weighted privacy |
|---|---|---|---|
| PPTD [15] | ✗ | ✓ | ✓ |
| L-PPTD [16] | ✗ | ✓ | ✓ |
| $L^2 - PPTD$ [16] | ✓ | ✓ | ✗ |
| EPTD [17] | ✗ | ✓ | ✓ |
| Scheme 1 [18] | ✗ | ✓ | ✓ |
| Scheme 2 [18] | ✗ | ✓ | ✓ |
| Scheme [26] | ✗ | ✓ | ✓ |
| Scheme [27] | ✗ | ✓ | ✓ |
| LPTD [41] | ✗ | ✓ | ✓ |
| RPTD-I | ✗ | ✓ | ✓ |
| RPTD-II | ✓ | ✓ | ✓ |

dual server. However, in the single-server-based scheme, the scheme requires other users to recalculate the perturbation values when some users leave, which may add additional computational and communication overhead for the users. In the dual-server-based scheme, users still need to be involved in the iterative process. Inspired by $L^2 - PPTD$, Zheng et al. [19] proposed a true value discovery scheme in ciphertext form using garbled circuits [25] and homomorphic encryption algorithms, and although this scheme also does not require user participation, the server needs to consume a larger computational overhead to generate the garbled circuits. To further improve the efficiency, some scholars have now also proposed differential privacy-based true value discovery schemes, e.g., Li et al. proposed a perturbation value-based scheme [26] and a local differential privacy-based scheme [27]. Through the analysis of both real and synthetic datasets, both schemes are shown to achieve a better balance between data privacy and data availability. In addition, we summarize data privacy preservation schemes that can be applied to mobile swarm intelligence awareness [28–33]. In general, representative techniques include (1) user anonymization [34, 35], (2) data scrambling [3, 36], and (3) cryptographic algorithm encryption [37–40]. A functional and privacy comparison between RPTD and existing schemes is given in Table 5.1.

## 5.1.3   Preliminary

### 5.1.3.1   Polynomial Function

Given a polynomial function $f(x) = a_n x^n + a_{n-1} x^{n-1} + \cdots + a_1 x + a_0$ with the highest power of $n$, where the coefficient of $x^i$ is $a_i$, $i \in [0, n]$, and $a_n \neq 0$, according to the basic theorem of algebra,

$$f(x) = a_n x^n + a_{n-1} x^{n-1} + \cdots + a_1 x + a_0$$
$$= a_n (x - x_n)(x - x_{n-1}) \cdots (x - x_1),$$

(5.1)

where $(x_1, x_2, \cdots, x_n)$ is the root of polynomial function $f(x)$. According to Weida's theorem, the $n - k$th coefficient of $f(x)$, i.e., $a_{n-k} (k \in [0, n])$, can be calculated as

$$\sum_{1 \leq i_1 < i_2 < \cdots < i_k \leq n} \left( \prod_{j=1}^{k} x_{i_j} \right) = (-1)^k \frac{a_{n-k}}{a_n}.$$

(5.2)

In the privacy-preserving task allocation scheme based on task contents designed in this book, a polynomial function is used to judge whether a task $x_i$ belongs to the task set $\{x_1, x_2, \cdots, x_n\}$. That is, suppose $\{\texttt{Task}(m) \rightarrow x_m\}_{m=1}^{n}$ is the required task set, and $f(x) = a_n (x - x_n)(x - x_{n-1}) \cdots (x - x_1)$ is the corresponding polynomial function. If $f(x_t) = 0$, the task $\texttt{Task}(m) \rightarrow x_t$ is in the required task set $\{\texttt{Task}(m) \rightarrow x_m\}_{m=1}^{n}$.

### 5.1.3.2   Searchable Encryption

Searchable encryption (SE) enables the third-party platform to find data or users that meet the requirements of data requesters in the form of ciphertext. From the perspective of encryption algorithms used, searchable encryption can be roughly divided into symmetric searchable encryption and public-key searchable encryption. This part mainly introduces the framework of searchable encryption scheme. On the whole, the searchable encryption scheme includes three entities: third-party platform, data holder, and data requester. Its basic process is summarized as follows:

- System Initialization: the data holder generates the key $sk$ and shares it with the data requester.
- Data Encryption: the data holder encrypts the data $w$ with the key and sends the ciphertext $E[w]$ to the third-party platform.
- Trapdoor Generation: for the search keyword $w'$, the data requester uses the key to generate $T[w']$ and sends it to the third-party platform.
- Data Retrieval: the third-party platform calculates the ciphertext $E[w]$ and trapdoor $T[w']$. If $w = w'$, output 1; otherwise, output 0.

If the SE scheme outputs 1 for any $w = w'$, then the SE scheme is correct. This book uses the SE framework to design a privacy-preserving task allocation scheme.

### 5.1.3.3 Truth Discovery

A truth discovery algorithm is an effective technology to calculate the real results of each task from multi-task data. On the whole, truth discovery mainly includes two steps: weight update and truth update. Suppose there are $K$ users and $M$ perception tasks, the perception data of user $u_k$ for task $o_m$ is $x^k_m$, the weight of user $u_k$ is $w_k$, and the truth of task $o_m$ is $x^*_m$. The following two steps are iterated until the iteration ends, and the truth of each task can be obtained:

- Weight Update: this step calculates the weight of the user according to the user data and the task truth. Specifically, the user's weight can be calculated as

$$w_k = f\left(\sum_{m=1}^{M} d\left(x^k_m, x^*_m\right)\right),\qquad(5.3)$$

  where $f(\cdot)$ is a monotonically decreasing function and $d(x^k_m, x^*_m)$ represents the distance between user data and truth.
- Truth Update: this step calculates the truth of the task according to the user's weight and data. Specifically, the truth can be calculated as

$$x^*_m = \frac{\sum_{k=1}^{K} w_k \cdot x^k_m}{\sum_{k=1}^{K} w_k}.\qquad(5.4)$$

In weight updating, the distance calculation function $d(\cdot)$ is slightly different according to different data types. For continuous data (such as speed, temperature, humidity, etc.), the square distance function $d(x^k_m, x^*_m) = (x^k_m - x^*_m)^2$ is used to represent the distance between user data and truth; for discrete data (such as question and answer options), assume that task $om$ has several answer options, $x^k_m = (0, \cdots, \frac{1}{q}, \cdots, 0)^T$ represents the q-th answer selected by user $u_k$, and $d(x^k_m, x^*_m) = (x^k_m - x^*_m)^T (x^k_m - x^*_m)$ represents the distance between user data and truth. Consistent with most of the existing truth discovery schemes [8, 21], this book selects the logarithmic equation to calculate the user's weight, i.e.,

$$w_k = \log\left(\frac{\sum_{k=1}^{K} \sum_{m=1}^{M} d\left(x^k_m, x^*_m\right)}{\sum_{m=1}^{M} d\left(x^k_m, x^*_m\right)}\right).\qquad(5.5)$$

In the privacy protection truth discovery scheme of truth disclosure and truth hiding designed in this book, the truth discovery algorithm is used to calculate the truth of the task.

### 5.1.3.4  Public-Key Cryptosystem Supporting Distributed Decryption

Liu et al. [42] proposed the public-key cryptosystem supporting distributed decryption (PCDD) based on the public-key encryption algorithm [43]. PCDD splits the decrypted private key and distributes it to several entities. Multiple entities decrypt ciphertext through cooperation. Because of its high efficiency and scalability, PCDD is widely used in secure multi-party computing and other applications. PCDD mainly includes the following algorithms:

- **KeyGen**($\kappa$)(Initialization): given a security parameter $\kappa$ select two large prime numbers $p, q, |p| = |q| = \kappa$, and calculate $n = pq$, $\lambda = lcm(p-1)(q-1)/2$, where $lcm(a, b)$ is the least common multiple of $a$ and $b$. Select random number $g \in Z^*_{n^2}$ and random number $\theta \in [1, n/4]$, and calculate $h = g^\theta \bmod n^2$. The public key is $pk = (n, g, h)$, and the weak private key is $\theta$. The strong private key is $\lambda$.
- **Enc**($pk, m$)(Data Encryption): given the data $m \in Z_n$, the algorithm encrypts $m$ as $\mathbf{E}[m] = h^r(1 + nm) \bmod n^2, r \in [1, n/4]$.
- **SDec**($\lambda, \mathbf{E}[m]$)(Ciphertext Decryption): given the ciphertext $\mathbf{E}[m]$, the algorithm uses a strong private key to decrypt the ciphertext $m = L((\mathbf{E}[m])^\lambda \bmod n^2)\lambda^{-1} \bmod n$, where $(\mathbf{E}[m])^\lambda \bmod n^2 = g^{\theta r\lambda}(1+mn\lambda) \bmod n^2$, $L(x) = \frac{x-1}{n}$.
- **SkeyS**($\lambda, t$)(Private Key Decomposition): given the strong private key $\lambda$, the algorithm divides $\lambda$ into t parts $(\lambda_1, \lambda_2, \cdots, \lambda_t)$, where $\sum_{i=1}^{t} \lambda_i \equiv 0 \bmod \lambda$, $\sum_{i=1}^{t} \lambda_i \equiv 1 \bmod n^2$.
- **PSDec**($\lambda, \mathbf{E}[m]$)(Partial Decryption): given the ciphertext $\mathbf{E}[m]$, the algorithm partially decrypts the ciphertext as $CT^{(i)} = (\mathbf{E}[m])_i^\lambda = g^{\theta r\lambda_i}(1+mn\lambda_i) \bmod n^2$.
- **DDec**($\{CT^{(i)}\}_{i=1}^{t}$)(Complete Decryption): given the partially decrypted ciphertext $CT^{(1)}, CT^{(2)}, \cdots, CT^{(t)}$, the algorithm performs complete decryption $m = L(\prod_{i=1}^{t} CT^{(i)})$.

For any $m \in Z_n$, $(\mathbf{E}[m])^{n-1} = h^{r(n-1)} \cdot (1 + (n-1)mn) \bmod n^2 = \mathbf{E}[-m]$. For the convenience of narration, this book uses $\mathbf{E}[m]^{-1}$ to represent $\mathbf{E}[m]^{n-1}$. PCDD has a homomorphic property similar to Paillier encryption system, that is, given plaintext $m_1, m_2, a \in Z_n$, PCDD satisfies

$$\mathrm{E}[m_1] \cdot \mathrm{E}[m_2] = \mathrm{E}[m_1 + m_2]$$
$$(\mathrm{E}[m])^a = \mathrm{E}[a \cdot m]. \tag{5.6}$$

PCDD is used to protect data privacy and construct privacy protection truth discovery algorithm.

## 5.2  Architecture Overview

This section describes the system model, security model, and design goals for RPTD.

### 5.2.1  System Model

The system model of RPTD consists of two main entities: the sensing user and the mobile crowdsensing platform. The functions of each entity are defined as follows:

- Mobile crowdsensing platform: The mobile crowdsensing platform is responsible for initializing the task information and the key information required by the system. RPTD uses two non-cooperative cloud servers (e.g., Amazon Cloud and Microsoft Cloud) as the mobile crowdsensing platform.
- Sensing user: The sensing user receives task information from the mobile crowdsensing platform and collects relevant sensing data using, for example, a mobile device. To protect the privacy of the data, the sensing user usually encrypts the data and uploads it to the mobile crowdsensing platform. It should be noted that perception users usually have poor computing and communication resources due to their environment and limited access to devices.

The problem to be solved in this chapter can be defined as follows: suppose now there are $K$ sensing users $\{u_1, u_2, \ldots, u_K\}$ and $M$ tasks $\{o_1, o_2, \ldots, o_M\}$, the sensing data of user $u_k$ for task $o_m$ is $x_m^k$, the weight of user $u_k$ is $\omega_k$, and the truth value of task $o_m$ is $x_m^*$. The purpose of RPTD is to obtain the true value $\{x_m^*\}_{m=1}^{M}$ from the data $\{x_m^k\}_{k,m=1}^{K,M}$ by truth discovery without disclosing the privacy of the sensing data and weights of the sensing users.

### 5.2.2  Security Model

In RPTD, we assume that both the cloud server and the sensing user are semi-trustworthy, i.e., both the cloud server and the sensing user execute the designed protocol honestly, but at the same time they are both curious about the privacy of others.

In particular, we assume that there is no collusion between the two cloud servers and the cloud server does not collude with the user, which is consistent with existing dual-server-based truth discovery schemes [16]. This assumption is reasonable in practical applications because (1) cloud servers are usually operated by large companies or organizations, and colluding with sensing users may greatly damage the reputation of the company or organization, and (2) due to geographic location and other factors, sensing users do not know each other, and in addition, even if sensing users know each other, they usually do not choose to collude with others due to their privacy concerns. In fact, there are already some schemes that can be used to resist collusion between cloud servers or between cloud servers and subscribers. Furthermore, there are already some schemes that can be used to resist collusion attacks between cloud servers or between cloud servers and subscribers, such as schemes [29, 44, 45], etc. Our scheme can be extended from these schemes, and the reader can consult the related literature for more technical details. In addition, this chapter mainly considers how to achieve privacy-preserving truth discovery, and thus witch attacks [46, 47], false data injection attacks [48–50], and external attacks [51] are not considered in this chapter.

## 5.2.3   Design Goals

The ultimate goal of RPTD is to achieve efficient and privacy-preserving true value discovery. Specifically, RPTD needs to achieve the following privacy and efficiency goals:

- Privacy: RPTD needs to protect the privacy of the data and weights of the sensing user, i.e., the cloud server or the sensing user cannot get the privacy information of other users from the existing data.
- Efficiency: RPTD should ensure that the computational and communication overhead at the perceived subscriber's end is as small as possible. In particular, for scenarios such as frequent user movement or poor user network conditions, RPTD should allow the user to not have to participate in the iterative process of truth discovery.

## 5.3   Detailed Design

This section describes the detailed design flow of RPTD. Specifically, this section first introduces the privacy-preserving truth discovery scheme RPTD-I where users participate in the truth discovery iterative process, and then it introduces the privacy-preserving truth discovery scheme RPTD-II where users do not participate in the truth discovery iterative process.

For the convenience of description, this section uses $S_A$ to represent a cloud server, $S_B$ to represent another cloud server, and PCDD to represent a public-key cryptosystem that supports multi-party collaborative decryption, $E[x]$ is used to represent the result obtained after encrypting $x$ with PCDD encryption algorithm $Enc(pk, x)$. $D[x]$ to represent the result obtained after decrypting x using PCDD decryption algorithm $SDec(\lambda, x)$. The specific descriptions of the above-related algorithms have been given in Section 5.1.3.4 and will not be repeated in this chapter.

## 5.3.1 RPTD-I: A Scheme for User Participation in the Iterative Process of Truth Discovery

### 5.3.1.1 System Initialization

Given a security parameter $\kappa$, the cloud server $S_B$ runs the $KeyGen(\kappa)$ algorithm of PCDD to generate the public key $pk = (n, g, h)$ and the decryption key $sk = \lambda$. Subsequently, assuming that the sum of all user weights for any task is less than a constant $Q$, i.e., $\max\left(\sum_{k=1}^{K} \omega_k \cdot x_m^k\right) < Q$, $S_B$ generates a superlinear sequence $\mathbf{a} = [a_1, a_2, \ldots, a_M]$, where $a_1 > Q$, $\sum_{m=1}^{j-1} a_m \cdot Q < a_j$, $j = 2, 3, \ldots, M$, $\sum_{m=1}^{M} a_m \cdot Q < n$. When the above operation is completed, $S_B$ saves the decrypted private key $\lambda$ and publishes the public key $pk$ and the superlinear sequence $\mathbf{a}$ to other entities.

### 5.3.1.2 The Iterative Process

The cloud server $S_A$ first initializes the true value $\{x_m^*\}_{m=1}^{M}$ for all tasks and exposes it to all aware users.

- Step I: When the true value is received, the sensing user $u_k$ first calculates the distance between each sensing data $\{x_m^k\}_{m=1}^{M}$ and its corresponding true value $\{x_m^*\}_{m=1}^{M}$, and then aggregates the distances of all tasks to obtain $sk = \sum_{m=1}^{M} d\left(x_m^k, x_m^*\right)$, and encrypts it as $E[sk]$. Finally, $u_k$ uploads the ciphertext to $S_A$, where $d\left(x_m^k, x_m^*\right)$ is the distance calculation function. In this chapter, only continuous data are considered; thus, $d\left(x_m^k, x_m^*\right) = \left(x_m^k, x_m^*\right)^2$.
- Step II: After receiving the ciphertext from the sensing user, $S_A$ aggregates all the ciphertexts to get $\sum_{k=1}^{K} E[sk] = E\left[\sum_{k=1}^{K} sk\right]$. Then, $S_A$ sends the aggregated ciphertexts to $S_B$.
- Step III: After receiving the aggregated ciphertext uploaded by the $S_A$, $S_B$ performs the decryption operation using the private key $sk$ to obtain $D\left[E\left[\sum_{K=1}^{K} sk\right]\right] = \sum_{k=1}^{K} sk$. Afterwards, $S_B$ sends $sk$ to all sensing users.
- Step IV: The sensing user $u_k$ calculates its own weight $\omega_K = \log\left(\frac{\sum_{k=1}^{K} sk}{sk}\right)$ based on the received aggregation distance. $u_k$ then aggregates the sensing data using a superlinear sequence, i.e., $u_k$ calculates

$$s_{\omega k} = \omega_k \cdot \left(a_1 \cdot x_1^k + a_2 \cdot x_2^k + \cdots + a_M \cdot x_M^k\right). \tag{5.7}$$

Subsequently, $u_k$ encrypts the weights and weighted sensory data and uploads $E[\omega_k]$, $E[s_{\omega k}]$ to $S_A$.

1:  Set $X_M = \sum_{k=1}^{K} s_{\omega k}, m = M$.
2:  Q1: If $m \geq 2$, calculate $X_{m-1} = X_m \bmod a_m$; Otherwise, terminate Q1 and Q2.
3:  Q2: Calculate $WX_m = \frac{X_m - X_{m-1}}{a_m}$, set $m = m - 1$, repeat Q1.
4:  $WX_1 = \frac{x_1}{a_1}$.
5:  Return $\{WX_m\}_{m=1}^{M}$.

**Fig. 5.1** The reduction process of aggregation weighted results for each task in RPTD-I

- Step V: $S_A$ aggregates the received user ciphertexts separately to obtain $\prod_{k=1}^{K} E[\omega_k]$ and $\prod_{k=1}^{K} E[s_{\omega k}]$. After that, $S_A$ sends the ciphertext to $S_B$.
- Step VI: $S_B$ decrypts the ciphertext to obtain $\sum_{k=1}^{K} \omega_k, \sum_{k=1}^{K} s_{\omega k}$, respectively. Afterwards, $S_B$ runs the reduction algorithm as in Fig. 5.1 to reduce the aggregated weighted result of each task, i.e., $WX_m = \sum_{k=1}^{K} \omega_k \cdot x_m^k$. $S_B$ computes $\frac{WX_m}{\sum_{k=1}^{K} \omega_k} = x_m^k$ to update the true value of task $o_m$ and sends it to all the sensing users.

Repeat Steps I to VI above, and the iteration is terminated when the difference between the two consecutive true values is less than a threshold or a set number of iterations is reached. At this point, the mobile crowdsensing platform can get the updated true value. It should be noted that RPTD-I only considers continuous data, and categorical data are not taken into account. In fact, categorical data can also be considered as special continuous data and can be applied in this scheme as well. In addition, RPTD-I uses superlinear sequences to aggregate the user's sensory data, reducing the computational and communication overhead for the user. If the number of users or tasks is particularly large, the sensing users can group the tasks and aggregate them separately using the superlinear sequence. Accordingly, the $S_A$ can also group the sensing users and aggregate the ciphertexts of the grouped users separately, thus preventing data overflow. In practice, the sensory data or weights may be fractional, in which case we use a factor L (usually L is a multiple of 10) to round up the fractional numbers. When the final result is obtained, the correct result is then reduced by dividing the factor L. The process of RPTD-I is shown in Fig. 5.2.

The correctness of RPTD-I is related to two aspects: (1) the correctness of the PCDD ciphertext aggregation and (2) the correctness of the reduction of the weighted results of each task aggregation. Next, we discuss and prove them separately.

We first prove the correctness of the PCDD ciphertext aggregation. Specifically, we need to prove that PCDD satisfies homomorphism, i.e., $E[m_1] \cdot E[m_2] = E[m_1 + m_2]$, and here we refer to the literature [51] to give the corresponding proof.

**Theorem 5.1** *The data $m_i$ is encrypted using the PCDD encryption algorithm to obtain the ciphertext $[m_i] = (1 + n \cdot m_i) \cdot h^{n \cdot r_i} \bmod n^2$. Given a series*

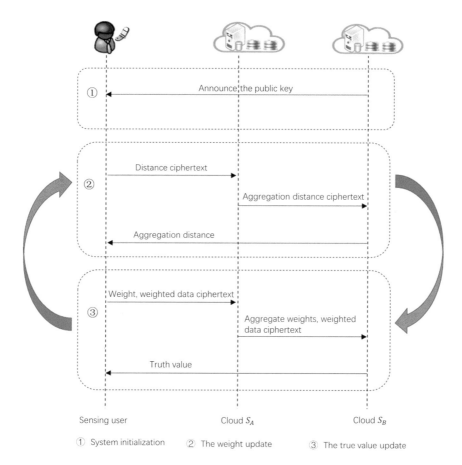

**Fig. 5.2**  RPTD-I system flow

of plaintexts $\{m_i\}_{i=1}^t$ of a ciphertext $\left\{[m_i] = (1 + n \cdot m_i) \cdot h^{n \cdot r_i} \bmod n^2\right\}_{i=1}^t$, the following equation holds:

$$\prod_{i=1}^{t}(1 + n \cdot m_i) \cdot h^{n \cdot r_i} \bmod n^2 = \left(1 + n \cdot \sum_{i=1}^{t} m_i\right) \cdot h^{n \cdot \sum_{i=1}^{t} r_i} \bmod n^2.$$

$$(5.8)$$

***Proof*** We prove this using mathematical induction. When $t = 1$, it is clear that the left side of the equation is equal to the right side of the equation. When $t = k$, assume that $\prod_{i=1}^{k}(1 + n \cdot m_i) \cdot h^{n \cdot r_i} \bmod n^2 = \left(1 + n \cdot \sum_{i=1}^{k} m_i\right) \cdot h^{n \cdot \sum_{i=1}^{k} r_i} \bmod n^2$ holds. Then it should be shown that the above equation still holds when $t = k + 1$. Specifically, when $t = k + 1$, the left side of the equation is

$$\prod_{i=1}^{k+1} (1 + n \cdot m_i) \cdot h^{n \cdot r_i} \bmod n^2$$

$$= \prod_{i=1}^{k} (1 + n \cdot m_i) \cdot h^{n \cdot r_i} \cdot (1 + n \cdot m_{k+1}) \cdot h^{n \cdot r_{k+1}} \bmod n^2$$

$$= \left(1 + n \cdot \sum_{i=1}^{k} m_i\right) \cdot h^{n \cdot \sum_{i=1}^{k} r_i} \cdot (1 + n \cdot m_{k+1}) \cdot h^{n \cdot r_{k+1}} \bmod n^2$$

$$= \left(1 + n \cdot \sum_{i=1}^{k+1} m_i\right) \cdot h^{n \cdot \sum_{i=1}^{k+1} r_i} \bmod n^2. \tag{5.9}$$

Since the left side of the equation is equal to the right side of the equation, it is known that the PCDD algorithm satisfies $E[m_1] \cdot E[m_2] = E[m_1 + m_2]$, which proves the correctness of the ciphertext aggregation. □

We next analyze the correctness of the reduction of the aggregated weighted results for each task. Specifically, from Fig. 5.1, we know that $X_M = \sum_{k=1}^{K} s_{\omega k}$, i.e.,

$$X_M = a_1 \sum_{k=1}^{K} \omega_k x_1^k + a_2 \sum_{k=1}^{K} \omega_k x_2^k + \cdots + a_{M-1} \sum_{k=1}^{K} \omega_k x_{M-1}^k + a_M \sum_{k=1}^{K} \omega_k x_M^k. \tag{5.10}$$

Since the aggregated weighted result for each task is assumed to be less than the constant Q, it follows that

$$a_1 \sum_{k=1}^{K} \omega_k x_1^k + a_2 \sum_{k=1}^{K} \omega_k x_2^k + \cdots + a_{M-1} \sum_{k=1}^{K} \omega_k x_{M-1}^k$$

$$< a_1 \cdot Q + a_2 \cdot Q + \cdots + a_{m-1} \cdot Q \tag{5.11}$$

$$= \sum_{i=1}^{m-1} a_i \cdot Q < a_m.$$

Therefore, $X_{m-1} = X_m \bmod a_m = a_1 \sum_{k=1}^{K} \omega_k x_1^k + a_2 \sum_{k=1}^{K} \omega_k x_2^k + \cdots + a_{m-1} \sum_{k=1}^{K} \omega_k x_{m-1}^k$, which leads to

$$\frac{X_m - X_{m-1}}{a_m} = \frac{a_m \sum_{k=1}^{K} \omega_k x_m^k}{a_m} = \sum_{k=1}^{K} \omega_k x_m^k. \tag{5.12}$$

Thus, the correctness of the reduction of the aggregated weighted results can be known.

In summary, the correctness of RPTD-I is confirmed.

## 5.3.2  RPTD-II: A Scheme Where Users Do Not Need to Participate in the Truth Discovery Iterative Process

Although RPTD-I can efficiently perform privacy-preserving truth discovery algorithms and protect the privacy of sensing user data and weights, the scheme still faces several problems. (1) The user still needs to participate in the iterative truth discovery process, which is less applicable in some mobile scenarios or when the sensing user has a poor network environment and weak computational communication capabilities. (2) Sensing users compute the true values themselves, and some selfish users may change their own weights to gain more benefits. In order to prevent the sensing users from tampering with their own weights and to further reduce the computational and communication overhead of the sensing users, this section proposes a privacy-preserving truth discovery scheme RPTD-II based on the homomorphism of the PCDD algorithm and the interaction between two cloud servers without the users' participation in the truth discovery iterative process. Similar to RPTD-I, RPTD-II also contains two parts: system initialization and iterative process.

### 5.3.2.1  System Initialization

- Step Q1: Given a security parameter $\kappa$, the cloud server $S_B$ runs PCDD's $KeyGen(\kappa)$ algorithm to generate a public key $pk = (n, g, h)$ and a decryption key $sk = \lambda$. $S_B$ saves the private key $\lambda$ and exposes the public key to all sensing users and the cloud server $S_A$.
- Step Q2: Sensing user $u_k$ selects random number $\alpha_m^k$ to disturb the sensing data $\tilde{x}_m^k = x_m^k - \alpha_m^k$. Then, $u_k$ encrypts the sum of random number and random number to obtain $\{E[\alpha_m^k]\}_{m=1}^M$ and $E\left[\sum_{m=1}^M (\alpha_m^k)^2\right]$. Finally, $u_k$ sends $\{\tilde{x}_m^k, E[\alpha_m^k]\}_{m=1}^M$ and $E\left[\sum_{m=1}^M (\alpha_m^k)^2\right]$ to $S_A$ and random number $\{\alpha_m^k\}_{m=1}^M$ to $S_B$.

The preceding steps are performed only once in the RPTD-II solution. Similar to RPTD-I, for non-integer perception data or random numbers, we use a coefficient $L$ (usually a multiple of 10) to round them up. When the final result is obtained, the correct result is restored by dividing the coefficient $L$.

### 5.3.2.2  The Iterative Process

$S_A$ first initializes the true value $\{x_m^*\}_{m=1}^M$ for all tasks and collaborates with $S_B$ to perform the following operations to run the truth discovery algorithm for privacy protection:

- Step I: SA first computes the distance between the true value and the sensing data. Specifically, for each sensing user $u_k$, $S_A$ calculates

$$
\begin{aligned}
C_{distcon}^k &= E\left[\sum_{m=1}^M \left(\alpha_m^k\right)^2\right] \cdot \prod_{m=1}^M E\left[\alpha_m^k\right]^{2 \cdot \left(\tilde{x}_m^k - x_m^*\right)} \\
&= E\left[\sum_{m=1}^M \left(2\alpha_m^k \cdot \left(\tilde{x}_m^k - x_m^*\right) + \left(\alpha_m^k\right)^2\right)\right].
\end{aligned}
\tag{5.13}
$$

Then, $S_A$ aggregates the ciphertext of all users and selects a random number $B$ to calculate $\left(\prod_{k=1}^K C_{distcon}^k\right)^B$. Similarly, $S_A$ selects a random number $b_k$ to calculate $\left(C_{distcon}^k\right)^{b_k}$ for each user's encrypted distance. After completing the above operations, $S_A$ sends $\left\{\sum_{k=1}^K \sum_{m=1}^M \left(\tilde{x}_m^k - x_m^*\right)^2, \left(\prod_{k=1}^K C_{distcon}^k\right)^B\right\}$, and $\left\{b_k \sum_{m=1}^M \left(\tilde{x}_m^k - x_m^*\right)^2, \left(C_{distcon}^k\right)^{b_k}\right\}_{k=1}^K$ to the cloud $S_B$. In particular, we set $B > \max(b_k)$.

- Step II: When receiving the ciphertext, $S_B$ uses the private key $\lambda$ to decrypt the encrypted distance of the user and calculates the distance of each sensing user as

$$
\begin{aligned}
sum_{dk} &= b_k \sum_{m=1}^M \left(\tilde{x}_m^k - x_m^*\right)^2 + D\left[\left(C_{distcon}^k\right)^{b_k}\right] \\
&= b_k \sum_{m=1}^M \left(\tilde{x}_m^k - x_m^*\right)^2 + b_k \sum_{m=1}^M \left(2\alpha_m^k \cdot \left(\tilde{s}_m^k - x_m^*\right) + \left(]\alpha_m^k\right)^2\right) \\
&= b_k \sum_{m=1}^M \left(x_m^k - x_m^*\right)^2.
\end{aligned}
\tag{5.14}
$$

Similarly, $S_B$ calculates the aggregation encryption distance for all users:

$$sum_d = B \sum_{m=1}^{M} \left( \tilde{s}_m^k - x_m^* \right)^2 + D \left[ \left( C_{distcon}^k \right)^B \right]$$

$$= B \sum_{m=1}^{M} \left( \tilde{s}_m^k - x_m^* \right)^2 + B \sum_{m=1}^{M} \left( 2\alpha_m^k \cdot \left( \tilde{s}_m^k - x_m^* \right) + \left( \alpha_m^k \right)^2 \right) \quad (5.15)$$

$$= B \sum_{m=1}^{M} \left( x_m^k - x_m^* \right)^2 .$$

Based on the above two formulas, the weight of $u_k$ can be calculated as

$$\tilde{\omega}_k = \log \left( B \sum_{k=1}^{K} \sum_{m=1}^{M} \left( x_m^k - x_m^* \right)^2 \right) - \log \left( b_k \sum_{m=1}^{M} \left( x_m^k - x_m^* \right)^2 \right) \quad (5.16)$$

$$= \omega_k + \log B_k$$

where $B_k = \frac{B}{b_k}$.

In order to protect the privacy of weights, $S_B$ selects a random number $c_k$ to disturb the weights as $\tilde{\omega}_k + c_k$. Finally, $S_B$ encrypts the random number to get $E[c_k]$ and returns $\{\tilde{\omega}_k + c_k, E[c_k]\}_{k=1}^{K}$ to $S_A$.

- Step III: When receiving the weight of the disturbance from $S_B$, $S_A$ first computes $\tilde{\omega}_k + c_k - \log B_k = \omega_k + c_k$, and then

$$\widetilde{WX}_m = \sum_{k=1}^{K} \left( x_m^k - \alpha_m^k \right) (\omega_k + c_k) . \quad (5.17)$$

To eliminate the random numbers $\alpha_m^k$ and $c_k$, $S_A$ selects the random numbers $\beta_1^m$ and $\beta_2^m$ to compute the ciphertext

$$C_c^m = E \left[ \beta_1^m \right] \cdot \prod_{k=1}^{K} E [c_k]^{x_m^k - \alpha_m^k} .$$

$$= E \left[ \beta_1^m + \sum_{k=1}^{K} \left( c_k \cdot \left( x_m^k - \alpha_m^k \right) \right) \right] . \quad (5.18)$$

$$C_c^m = E\left[\beta_2^m\right] \cdot \prod_{k=1}^{K} E\left[\alpha_m^k\right]^{\omega_k + c_k}.$$

$$= E\left[\beta_2^m + \sum_{k=1}^{K}\left(\alpha_m^k \cdot (\omega_k + c_k)\right)\right]. \tag{5.19}$$

The ciphertext M is returned to $S_B$.

- Step IV: $S_B$ uses the private key to decrypt the received ciphertext and obtain

$$\text{sum}_{rand}^m = D\left[C_c^m\right] + \sum_{k=1}^{K} c_k \cdot \alpha_m^k - D\left[C_\alpha^m\right]$$

$$= \sum_{k=1}^{K}\left(\left(x_m^k - \alpha_m^k\right) \cdot c_k + c_k \cdot \alpha_m^k - \alpha_m^k \cdot (w_k + c_k)\right) + \beta_1^m - \beta_2^m. \tag{5.20}$$

Finally, $S_B$ sends $\left\{\text{sum}_{rand}^m\right\}_{m=1}^{M}$ and $\sum_{k=1}^{K} c_k$ to $S_A$.

- Step V: When receiving data from $S_B$, $S_A$ updates the true value of task $o_m$ according to the following formula:

$$x_m^* = \frac{\widetilde{WX}_m - \text{sum}_{rand}^m + \beta_1^m - \beta_2^m}{\sum_{k=1}^{K}(w_k + c_k) - \sum_{k=1}^{K} c_k}. \tag{5.21}$$

Similar to RPTD-I, Steps I–V are iterated until the iteration is terminated and the updated true value is obtained. Since the iterative process is done by two cloud servers, the sensing user does not need to participate in the truth discovery iteration process, thus saving the user a lot of computational and communication overhead. The detailed process of RPTD-II is shown in Fig. 5.3.

## 5.4  Security Analysis

In this section, we analyze the security and privacy of RPTD-I and RPTD-II. Before starting the formal analysis, we first give a formal definition of security for the semi-plausible adversary model based on the literature [44, 52].

**Definition 5.1** Suppose that protocol $P$ requires $A$ to compute $f_A(x, y)$ and $B$ to compute $f_B(x, y)$, where $x, y$ are the inputs of $A, B$, respectively. Let $view_A(x, y)$, $view_A(x, y)$ be the results observed by $A$ and $B$ through protocol

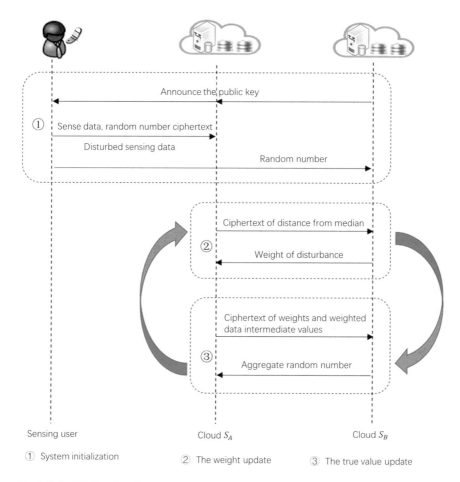

Fig. 5.3  RPTD-II system flow

$P$ for $(x, y)$. That is, if $(x, r_A)$ (or $(y, r_B)$) is the input of $A$ (or $B$) and a random number, and $m_i$ is the i-th message uploaded to protocol $P$, then

$$\text{view}_{\mathcal{A}} = (x, r_{\mathcal{A}}, m_1, m_2, \cdots, m_t)$$
$$\text{view}_{\mathcal{B}} = (y, r_{\mathcal{B}}, m_1, m_2, \cdots, m_t). \tag{5.22}$$

Let $O_A(x, y)$, $O_B(x, y)$ be the outputs of $A$, $B$. If there exist polynomial time (P.P.T) simulators S1, S2 such that

$$(S_1(x, f_{\mathcal{A}}(x, y)), f_{\mathcal{B}}(x, y)) \equiv (\text{view}_{\mathcal{A}}(x, y), O_{\mathcal{B}}(x, y))$$
$$(f_{\mathcal{A}}(x, y), S_2(y, f_{\mathcal{B}}(x, y))) \equiv (O_{\mathcal{A}}(x, y), \text{view}_{\mathcal{B}}(x, y)) \tag{5.23}$$

where $\equiv$ indicates that the computation is indistinguishable, then protocol $P$ is resistant to semi-trustworthy adversaries.

It should be noted that the public-key cryptosystem PCDD, which supports multi-party cooperative decryption, used by RPTD has been shown to be semantically secure (i.e., the ciphertext is indistinguishable) in a semi-trusted model in scheme [42]. Based on the above definitions and conclusions, we give the following analysis.

### 5.4.1  RPTD-I Security and Privacy Analysis

**Theorem 5.2**  *If the public-key encryption algorithm PCDD is semantically secure, the number of sensing users participating in the system is greater than 3, and the product of the number of users and the number of tasks is greater than the number of truth discovery iterations, then in RPTD-I, the user or the mobile crowdsensing platform cannot be informed of the sensory data or weight information of other users through the existing data.*

**Proof**  For each sensing user $u_k$, the privacy information of $u_k$ is encrypted as $E[x]$, where $x$ represents the sensory data and weight information of the user. Since PCDD has been shown to resist selective plaintext attacks [42, 51], an attacker cannot restore the privacy information from the ciphertext without knowing the private key. Although $S_B$ has the private key, since the ciphertext has been aggregated at $S_A$, $S_B$ can only get the aggregated results $\sum_{k=1}^{K} s_k$, $\sum_{k=1}^{K} w_k$, $\sum_{k=1}^{K} \sum_{m=1}^{M} w_k \cdot x_m^k$ and cannot restore the sensitive information of individual users from these aggregated results. It is important to note that an attacker may try to restore individual information by constructing an equation, such as the cloud $S_B$ computed from the updated true value

$$\sum_{m=1}^{M} \left( x_m^{*,(1)} - x_m^1 \right)^2 + \sum_{m=1}^{M} \left( x_m^{*,(1)} - x_m^2 \right)^2 + \cdots + \sum_{m=1}^{M} \left( x_m^{*,(1)} - x_m^K \right)^2 = \sum_{k=1}^{K} s_k^{(1)}$$

$$\sum_{m=1}^{M} \left( x_m^{*,(2)} - x_m^1 \right)^2 + \sum_{m=1}^{M} \left( x_m^{*,(2)} - x_m^2 \right)^2 + \cdots + \sum_{m=1}^{M} \left( x_m^{*,(2)} - x_m^K \right)^2 = \sum_{k=1}^{K} s_k^{(2)}$$

$$\cdots$$

$$\sum_{m=1}^{M} \left( x_m^{*,(t)} - x_m^1 \right)^2 + \sum_{m=1}^{M} \left( x_m^{*,(t)} - x_m^2 \right)^2 + \cdots + \sum_{m=1}^{M} \left( x_m^{*,(t)} - x_m^K \right)^2 = \sum_{k=1}^{K} s_k^{(t)},$$

$$(5.24)$$

where $x_m^{*,(t)}$, $s_k^{(t)}$ represent the updated true value and the obtained aggregation distance at the $t$-th iteration, respectively. In the above equation, the number of unknowns is $M \cdot K$. Assuming that the number of iterations for truth discovery is $iter$, when $M \cdot K > iter$, the attacker cannot obtain the data information of each user because the number of unknowns is larger than the number of equations. When $iter = 10$, the true value discovery can be converged, and in the real mobile group intelligence perception scenario, $M \cdot K$ is usually much larger than $iter$, so it can be known that RPTD-I can effectively protect the user's privacy.

For users, $u_k$ knows the aggregated distance between $m$ and all users $\sum_{k=1}^{K} \sum_{m=1}^{M} d\left(x_m^k, x_m^*\right)$. Since the number of users in mobile group-wise perception is usually much larger than 3, $u_k$ cannot infer the data of other users from $\sum_{k=1}^{K} \sum_{m=1}^{M} d\left(x_m^k, x_m^*\right)$. $\qquad\square$

### 5.4.2   RPTD-II Security and Privacy Analysis

**Theorem 5.3** *If the public-key cryptography algorithm PCDD is semantically secure and the number of tasks participating in the system is greater than 2, then in RPTD-II, the mobile crowdsensing platform cannot be informed of the user's sensory data or weight information from the existing data.*

**Proof** In RPTD-II, since the sensing user does not need to participate in the truth discovery iterative process, we only need to analyze whether the cloud server is likely to get the sensory data or weights of the user based on the existing information.

For $S_A$, it knows the data including plaintext $x_m^k - \alpha_m^k$, $w_k + c_k$ and ciphertext $E\left[\alpha_m^k\right], E\left[\sum_{m=1}^{M}\left(\alpha_m^k\right)^2\right], C_{distcon}^k, C_c^m, C_\alpha^m$. Without knowing the random numbers $\alpha_m^k$ and $c_k$, $S_A$ cannot reduce the sensory data and weights from the plaintext information. Since PCDD has been proved to be semantically secure, $S_A$ cannot restore the plaintext information from the ciphertext without knowing the private key of PCDD.

For $S_B$, the data includes plaintext $B \sum_{k=1}^{K} \sum_{m=1}^{M}\left(\tilde{x}_m^k - x_m^*\right)^2$, $B \sum_{k=1}^{K} \sum_{m=1}^{M}\left(2\alpha_m^k \cdot \left(\tilde{x}_m^k - x_m^*\right)^2 + \left(\alpha_m^k\right)^2\right)$, $\left\{b_k \sum_{m=1}^{M}\left(\tilde{x}_m^k - x_m^*\right)^2, b_k \sum_{m=1}^{M}\left(2\alpha_m^k \cdot \left(\tilde{x}_m^k - x_m^*\right)^2 + \left(\alpha_m^k\right)^2\right)\right\}_{k=1}^{K}$, $\sum_{k=1}^{K}\left(\left(x_m^k - \alpha_m^k\right) \cdot c_k + c_k \cdot \alpha_m^k - \alpha_m^k \cdot \left(w_k + c_k\right)\right) + \beta_1^m - \beta_2^m$. Although $S_B$ knows the random number $\alpha_m^k$, $S_B$ cannot infer the user's sensory data and weights from it because it does not know the true value $x_m^*$, the random number $\beta_1^m$, $\beta_2^m$, $and$ $\log B_K$, and the number of tasks involved in the system is greater than 2. $S_B$ can also construct a series of equations to try to infer the sensing user's privacy information as illustrated in the proof of 5.2, but without knowing the true value $x_m^*$ and due to the introduction of the random number $B, b_k, \beta_1^m, \beta_2^m$, $S_B$ is unable to reduce the user's sensory data or weights from the available information. $\qquad\square$

## 5.5  Performance Evaluation and Analysis

In this section, a theoretical and experimental analysis of the performance of RPTD is given.

### 5.5.1  Theoretical Analysis

For the sake of description, we define ENC as the time required for one encryption operation of PCDD and DEC as the time required for one decryption operation of PCDD, and $|N|$ represents the size of one ciphertext obtained using the PCDD encryption algorithm. Moreover, we only consider the size of the amount of data sent by the entity, but not the size of the amount of data received by the entity. The size of the plaintext information, such as user aggregation distance and true value, is not considered since it is much smaller than the size of the ciphertext.

#### 5.5.1.1  Theoretical Analysis of RPTD-I

- **Computational overhead.** At the sensing user side, to get the aggregated distance of all users, each sensing user needs to compute the ciphertext $s_k$ with a computational overhead of 1 ENC. To compute the true value, the sensing user needs to encrypt the weight and weighted data, which requires 2 encryption operations and consumes 2 ENC. At the cloud server side, for each iteration, $S_A$ needs to perform $K - 1$ ciphertext. At the cloud server side, $S_A$ needs to perform $K - 1$ ciphertext multiplications to aggregate the encryption distance, weight, and weighted data of all sensing users for each iteration. Accordingly, $S_B$ needs to perform one ciphertext decryption operation to restore the plaintext data from the ciphertext, so the computational overhead of $S_B$ is 3 DEC per iteration.
- **Communication overhead.** At the sensing user side, in order to get the aggregated distance of all users, each user needs to upload the encrypted distance to $S_A$ with a communication overhead of $|N|$. After receiving the aggregated distance from all users, the sensing user needs to upload the weighted ciphertext and the aggregated weighted data ciphertext to $S_A$, with a communication overhead of $2|N|$. Therefore, the communication overhead of each sensing user during each iteration is $3|N|$. At the cloud server side, $S_A$ needs to interact with $S_B$ for each iteration. For $S_A$, it needs to send three ciphertexts to $S_B$ with a communication overhead of $3|N|$.

#### 5.5.1.2  Theoretical Analysis of RPTD-II

- **Computational overhead.** At the perceptive user side, the user only needs to send the data to two cloud servers without participating in the iterative process, and the total computational overhead is $(M + 1)$ ENC. At the cloud server side, $S_B$ needs to perform $K$ decryption operations to calculate the distance of each user, 1 decryption operation to calculate the aggregated distance of all users, $K$ encryption operations to encrypt the random numbers $c_k$, and $2M$ decryption operations to decrypt $C_c^m$, $C_\alpha^m$. $S_A$ needs to perform $2M$ encryption operations to encrypt the random numbers $\beta_1^m$, $\beta_2^m$. During each iteration, the computational overhead of the two cloud servers $S_A$, $S_B$ is $2M \cdot ENC$ and $K \cdot ENC + (K + 2M + 1) \cdot DEC$, respectively.
- **Communication overhead.** Each perceptive user needs to upload $M + 1$ cipher messages to $S_A$ in the initialization phase of the system. In each iteration, $S_A$ needs to send $K + 2M + 1$ ciphertexts to $S_B$ and $S_B$ needs to send $K + M$ ciphertexts to $S_A$. Although the user needs to upload multiple ciphertexts to $S_A$, the user does not need to participate in the iterative process of truth discovery, and still saves more communication overhead compared with the user-involved iterative scheme RPTD-I.

### 5.5.2  Performance Analysis

In this section, we experimentally verify the computational performance of RPTD. Specifically, we use a laptop with a 2.5 GHz Intel i7 processor and 16.0 GB RAM as the cloud server and an Android phone with 1.5 GHz and 2 GB RAM as the sensing user. All schemes are implemented in Java with security parameters set to 512 bits and encryption algorithms using PCDD, a public-key cryptosystem that supports multi-party cooperative decryption. For comparison, we implement PPTD using Paillier Threshold Encryption Toolbox 1 [15] and LPTD using the modified Paillier homomorphic encryption algorithm [51] to implement LPTD [41], and L-PPTD [16] using the Paillier homomorphic encryption algorithm to compare the computational efficiency of RPTD. Before comparing the schemes, we first analyze the correctness and convergence of RPTD.

- **Correctness analysis.** We first analyze the accuracy of RPTD by the true value discovery algorithm CRH [8] without privacy protection. Similar to scheme [16], we choose the root mean squared error (RMSE) of the true value and the computed true value as the evaluation criterion. Specifically, the number of observation tasks is set to 10, the number of users is from 3 to 10, and the data values are taken in the range [20, 30]. It can be seen from Fig. 5.4 that RPTD performs similarly to CRH, so it can be demonstrated that the cryptographic operation of RPTD does not sacrifice the functionality of truth discovery itself.

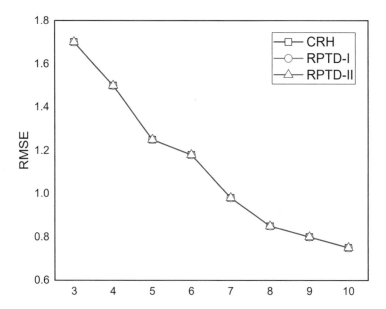

**Fig. 5.4** RPTD correctness analysis

- **Convergence analysis.** We next analyze the convergence of RPTD. The criterion for judging convergence is whether the difference between two consecutive true updates $\left| x^t - x^{t-1} \right|$ is less than a given threshold, where $x^t$ represents the truth value of the $t^{th}$ iteration and $x^0$ is randomly generated between [0, 1]. We choose a random number between [20, 30] as the truth value and generate a Gaussian function noise based on the true value to perturb it with the threshold set to 0.001. It can be seen from Fig. 5.5 that the convergence of the scheme is achieved when the number of iterations is 10.

### 5.5.2.1  Experiments Based on Real Crowdsensing Scenarios

We choose a real mobile swarm intelligence perception scenario, i.e., a reference scenario [53], using mobile devices to measure indoor floor plans. In this scenario, mobile sensors (e.g., gyroscope and compass) are used to collect the distance between two different locations. We use 10 cell phones as sensing users and select 25 corridor distances as the observed task with the number of iterations set to 10. Each experiment is run 10 times and the average value is chosen as the result of the experiment.

We calculate the required time on the perception user side and on the cloud server side separately. Table 5.2 gives the time spent on the perception user side when the number of observation tasks varies between 5 and 25. For example, when the number of observation tasks is 25, PPTD takes 0.219 s and LPTD takes

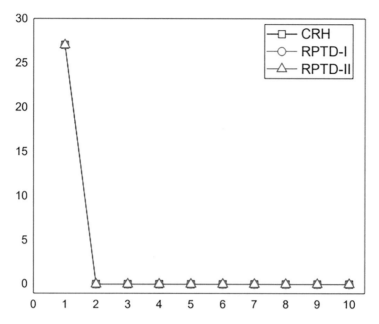

**Fig. 5.5** Convergence analysis of RPTD

0.051 s, while scheme RPTD-I takes only 0.033 s and scheme RPTD-II takes only 0.030 s during the whole system operation. Since L-PPTD does not encrypt the data at the user side, it has less computational overhead than the other schemes. The computational overhead of RPTD-I increases slightly with the number of sensing tasks, which we speculate is due to (1) the fact that RPTD-I requires data aggregation using superlinear sequences and the increase in sensing tasks leads to an increase in aggregation operation time, and (2) the instability of mobile device performance. Although the time of RPTD-I increases slightly in sensing the user side, its computational overhead is still smaller than schemes PPTD and LPTD.

**Table 5.2** Communication overhead of RPTD on the client side based on real crowdsensing scenarios

| Number of tasks | 5 | 10 | 15 | 20 | 25 |
|---|---|---|---|---|---|
| PPTD/iterations (sec) | 0.151 | 0.168 | 0.185 | 0.201 | 0.219 |
| LPTD/iterations (sec) | 0.013 | 0.023 | 0.032 | 0.041 | 0.051 |
| L-PPTD initialization ($10^{-2}$ s) | 0.005 | 0.009 | 0.010 | 0.016 | 0.024 |
| L-PPTD/iterations ($10^{-2}$ s) | 0.296 | 0.452 | 0.708 | 0.830 | 1.106 |
| RPTD-I/iterations (sec) | 0.024 | 0.022 | 0.026 | 0.030 | 0.033 |
| RPTD-II/initialization (sec) | 0.005 | 0.011 | 0.016 | 0.023 | 0.030 |

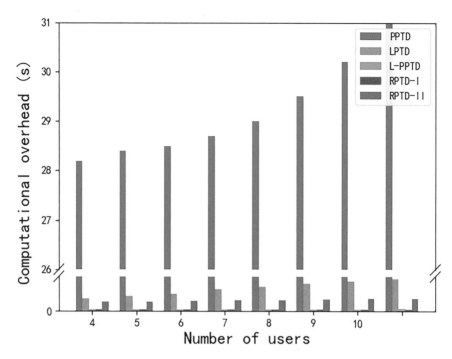

**Fig. 5.6** Computational overhead for different scenarios with different number of users

When the number of sensing users is from 3 to 10 and the number of observation tasks is fixed at 10, the computational overhead on the cloud server side during the truth discovery iteration is shown in Figs. 5.6 and 5.7. PPTD and LPTD consume more computational overhead than RPTD-I. This is because PPTD and LPTD use time-consuming threshold decryption algorithms and Paillier homomorphic encryption algorithms, while RPTD-I uses techniques such as superlinear sequences, data perturbation, etc. to improve computational efficiency. Similarly, RPTD-I also achieves higher computational efficiency compared to L-PPTD. RPTD-II takes more computation time than RPTD-I in the iterative process because RPTD-II requires additional computation of distance ciphertext and weight ciphertext for each sensing user. Although there is a sacrifice in computational efficiency, RPTD-II gives full protection to both the data and weights of the sensing users.

When the number of tasks is from 5 to 25, and the number of sensing users is fixed at 10, the computational overhead on the cloud server side for the truth discovery iteration process is shown in Figs. 5.8 and 5.9. The computational overhead of PPTD, LPTD, and L-PPTD increases with the number of tasks, while the computational overhead of RPTD-I remains almost constant. This is because RPTD-I uses a superlinear sequence to aggregate the weighted sensory data, and thus the encryption and decryption operations of the data do not vary with the number of tasks. RPTD-II takes more time on the cloud server.

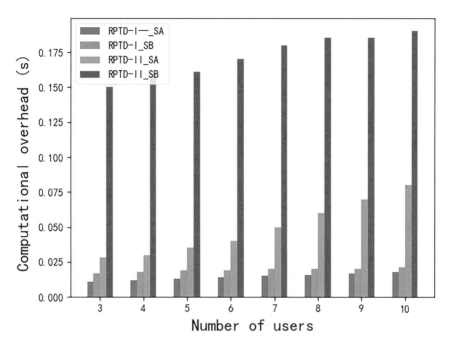

**Fig. 5.7** Computational overhead on the cloud server side for RPTD scenarios with different number of users

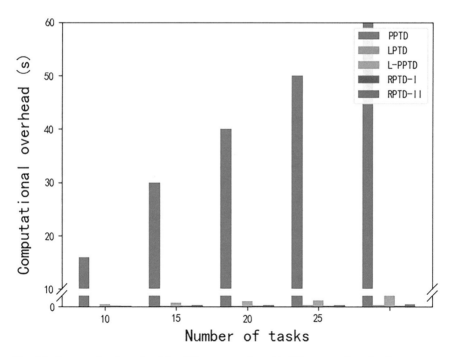

**Fig. 5.8** Computational overhead for different scenarios with different number of tasks

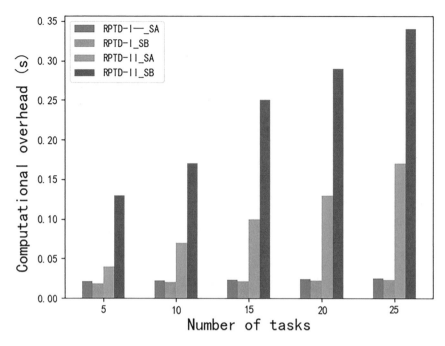

**Fig. 5.9** Computational overhead on the cloud server side for RPTD scenarios with different number of tasks

The above experiments on real mobile swarm intelligence perception scenarios validate that schemes RPTD-I and RPTD-II have high computational and communication efficiency when faced with real datasets.

### 5.5.2.2   Experiments Based on Simulated Crowdsensing Scenarios

To further demonstrate the performance of RPTD, we implemented all scenarios in a simulated mobile swarm intelligence perception scenario. Specifically, the scenario contains 1000 sensing users and 500 tasks. The sensory data were randomly generated by adding Gaussian noise to the original corridor distance dataset. We vary the number of sensing users and tasks separately to observe the computational overhead on the sensing user side and on the mobile crowdsensing platform. The number of iterations is set to 10, and each experiment is run 10 times and the average value is chosen as the experimental result.

Table 5.3 gives the computational overhead on the sensing user side when the number of tasks ranges from 100 to 500. It can be seen that the RPTD scheme is more efficient than the other schemes. For example, when the number of tasks is 500, PPTD takes 1.521 s, LPTD takes 0.946 s, and RPTD-I takes only 0.365 s. For RPTD-II, when the number of tasks is 500, the sensing user needs only 0.608 s to

**Table 5.3**  Computational overhead of RPTD on the client side based on simulated crowdsensing scenarios

| Number of tasks | 100 | 200 | 300 | 400 | 500 |
|---|---|---|---|---|---|
| PPTD/iterations (sec) | 0.415 | 0.689 | 0.939 | 1.234 | 1.521 |
| LPTD/iterations (sec) | 0.186 | 0.387 | 0.569 | 0.737 | 0.946 |
| L-PPTD initialization ($10^{-2}$ s) | 0.001 | 0.002 | 0.003 | 0.004 | 0.005 |
| L-PPTD/iterations ($10^{-2}$ s) | 0.044 | 0.088 | 0.127 | 0.170 | 0.211 |
| RPTD-I/iterations (sec) | 0.132 | 0.168 | 0.240 | 0.314 | 0.365 |
| RPTD-II/initialization (sec) | 0.120 | 0.239 | 0.360 | 0.503 | 0.608 |

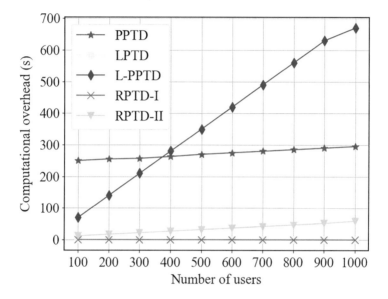

**Fig. 5.10**  The computational overhead of different scenarios with different number of users

complete the whole truth discovery process. We then observe the computational overhead on the cloud server side when the number of aware users and tasks changes.

As shown in Figs. 5.10 and 5.11, when the number of tasks is 100 and the number of sensing users is 100 to 1000, RPTD achieves a higher efficiency compared to the other three schemes.

As shown in Figs. 5.12 and 5.13, when the number of users is 1000 and the number of tasks is 100 to 500, RPTD achieves a higher efficiency compared to the other three schemes.

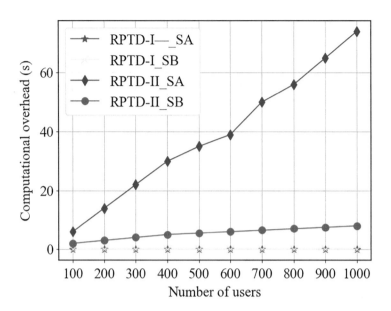

**Fig. 5.11** The computational overhead of RPTD scenarios on different cloud servers with different number of users

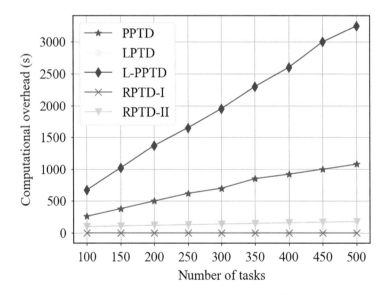

**Fig. 5.12** The computational overhead of different scenarios with different number of tasks

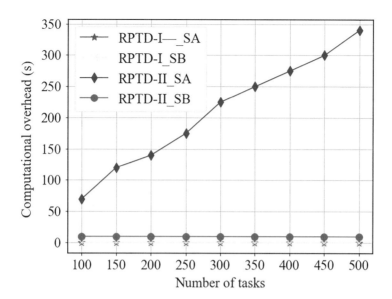

**Fig. 5.13** The computational overhead of RPTD scenarios on different cloud servers with different number of tasks

## 5.6 Summary

In this chapter, we propose two privacy-preserving truth discovery schemes, RPTD-I and RPTD-II, for the mobile group-wise perception scenario with truth disclosure. Among them, RPTD-I is suitable for scenarios where the user's location is relatively fixed and can participate in the truth discovery iterative process. With the homomorphic encryption algorithm PCDD and superlinear sequences, RPTD-I can significantly reduce the computational and communication overhead of users and cloud servers. RPTD-II is suitable for scenarios where users move frequently and cannot participate in the truth discovery iterative process. By using data perturbation techniques and exploiting the homomorphism of the cryptographic algorithm PCDD, RPTD-II allows two cloud servers to interactively and iteratively update the true values and weights in a privacy-preserving manner. The security and privacy analysis demonstrates that RPTD can effectively protect the user's sensory data and weight privacy. Based on this chapter, it is a meaningful research direction to investigate verifiable privacy-preserving true value discovery schemes considering malicious behaviors of the cloud servers.

# References

1. Gao, R., Zhao, M., Ye, T., Ye, F., Wang, Y., Bian, K., Wang, T., Li, X.: Jigsaw: indoor floor plan reconstruction via mobile crowdsensing. In: The 20th Annual International Conference on Mobile Computing and Networking (MobiCom), 17 Sep 2017 17:21:15 +0200, pp. 249–260 (2014)
2. Hu, S., Su, L., Liu, H., Wang, H., Abdelzaher, T.F.: Smartroad: smartphone-based crowd sensing for traffic regulator detection and identification. Trans. Sensor Netw. **11**(4), 55:1–55:27 (2015)
3. Zhang, C., Zhu, L., Xu, C., Lu, R.: PPDP: an efficient and privacy-preserving disease prediction scheme in cloud-based e-healthcare system. Futur. Gener. Comput. Syst. **79**, 16–25 (2018)
4. Mukherjee, S., Weikum, G., Danescu-Niculescu-Mizil, C.: People on drugs: credibility of user statements in health communities. In: Proceedings of the 20th ACM SIGKDD International Conference on Knowledge Discovery and Data Mining, 24 Sep 2014 13:01:29 +0200, pp. 65–74 (2014)
5. Baier, P., Dürr, F., Rothermel, K.: Efficient distribution of sensing queries in public sensing systems. In: IEEE 10th International Conference on Mobile Ad-Hoc and Sensor Systems, 15 Jun 2017 21:32:26 +0200, pp. 272–280 (2013)
6. Mao, X., Miao, X., He, Y., Li, X.-Y., Liu, Y.: Citysee: urban $CO2$ monitoring with sensors. In: IEEE Conference on Computer Communications (INFOCOM), 19 Mar 2018 16:28:26 +0100, pp. 1611–1619 (2012)
7. Li, Q., Li, Y., Gao, J., Su, L., Zhao, B., Demirbas, M., Fan, W., Han, J.: A confidence-aware approach for truth discovery on long-tail data. Proc. VLDB Endowment **8**(4), 425–436 (2014)
8. Li, Q., Li, Y., Gao, J., Zhao, B., Fan, W., Han, J.: Resolving conflicts in heterogeneous data by truth discovery and source reliability estimation. In: Proceedings of the 2014 ACM SIGMOD International Conference on Management of Data, pp. 1187–1198 (2014)
9. Li, Y., Li, Q., Gao, J., Su, L., Zhao, B., Fan, W., Han, J.: On the discovery of evolving truth. In: Proceedings of the 21th ACM SIGKDD International Conference on Knowledge Discovery and Data Mining, pp. 675–684 (2015)
10. Ma, F., Li, Y., Li, Q., Qiu, M., Gao, J., Zhi, S., Su, L., Zhao, B., Ji, H., Han, J.: Faitcrowd: fine grained truth discovery for crowdsourced data aggregation. In: Proceedings of the 21th ACM SIGKDD International Conference on Knowledge Discovery and Data Mining, pp. 745–754 (2015)
11. Meng, C., Jiang, W., Li, Y., Gao, J., Su, L., Ding, H., Cheng, Y.: Truth discovery on crowd sensing of correlated entities. In: Proceedings of the 13th ACM Conference on Embedded Networked Sensor Systems, pp. 169–182 (2015)
12. Yin, X., Han, J., Philip, S.Y.: Truth discovery with multiple conflicting information providers on the web. IEEE Trans. Knowl. Data Eng. **20**(6), 796–808 (2008)
13. Zheng, Y., Li, G., Li, Y., Shan, C., Cheng, R.: Truth inference in crowdsourcing: Is the problem solved? Proc. VLDB Endowment **10**(5), 541–552 (2017)
14. Singla, A., Krause, A.: Incentives for privacy tradeoff in community sensing. In: Proceedings of the AAAI Conference on Human Computation and Crowdsourcing (2013)
15. Miao, C., Jiang, W., Su, L., Li, Y., Guo, S., Qin, Z., Xiao, H., Gao, J., Ren, K.: Cloud-enabled privacy-preserving truth discovery in crowd sensing systems. In: Proceedings of the 13th ACM Conference on Embedded Networked Sensor Systems, pp. 183–196 (2015)
16. Miao, C., Su, L., Jiang, W., Li, Y., Tian, M.: A lightweight privacy-preserving truth discovery framework for mobile crowd sensing systems. In: IEEE Conference on Computer Communications (INFOCOM), pp. 1–9 (2017)
17. Xu, G., Li, H., Tan, C., Liu, D., Dai, Y., Yang, K.: Achieving efficient and privacy-preserving truth discovery in crowd sensing systems. Comput. Secur. **69**, 114–126 (2017)
18. Zheng, Y., Duan, H., Yuan, X., Wang, C.: Privacy-aware and efficient mobile crowdsensing with truth discovery. IEEE Trans. Dependable Secure Comput. **PP**(99), 1–1 (2017)

19. Zheng, Y., Duan, H., Wang, C.: Learning the truth privately and confidently: encrypted confidence-aware truth discovery in mobile crowdsensing. IEEE Trans. Inf. Forensics Secur. **13**(10), 2475–2489 (2018)
20. Damgård, I., Jurik, M.: A generalisation, a simplification and some applications of paillier's probabilistic public-key system. In: International Workshop on Public Key Cryptography, pp. 119–136. Springer, Berlin (2001)
21. Li, Y., Li, Q., Gao, J., Su, L., Zhao, B., Fan, W., Han, J.: Conflicts to harmony: a framework for resolving conflicts in heterogeneous data by truth discovery. IEEE Trans. Knowl. Data Eng. **28**(8), 1986–1999 (2016)
22. Zhou, J., Cao, Z., Dong, X., Lin, X.: Security and privacy in cloud-assisted wireless wearable communications: challenges, solutions, and future directions. IEEE Wirel. Commun. **22**(2), 136–144 (2015)
23. Li, S., Xue, K., Yang, Q., Hong, P.: PPMA: privacy-preserving multi-subset aggregation in smart grid. IEEE Trans. Ind. Inf. **14**(2), 462–471 (2018)
24. Lu, R., Liang, X., Li, X., Lin, X., Shen, X.: EPPA: an efficient and privacy-preserving aggregation scheme for secure smart grid communications. IEEE Trans. Parallel Distrib. Syst. **23**(9), 1621–1631 (2012)
25. Yao, A.C.-C.: How to generate and exchange secrets. In: 27th Annual Symposium on Foundations of Computer Science, pp. 162–167 (1986)
26. Li, Y., Miao, C., Su, L., Gao, J., Li, Q., Ding, B., Qin, Z., Ren, K.: An efficient two-layer mechanism for privacy-preserving truth discovery. In: Proceedings of the 24th ACM SIGKDD International Conference on Knowledge Discovery & Data Mining (KDD), pp. 1705–1714 (2018)
27. Li, Y., Xiao, H., Qin, Z., Miao, C., Su, L., Gao, J., Ren, K., Ding, B.: Towards differentially private truth discovery for crowd sensing systems. CoRR, abs/1810.04760 (2018)
28. Du, X., Guizani, M., Xiao, Y., Chen, H.-H.: Secure and efficient time synchronization in heterogeneous sensor networks. IEEE Trans. Veh. Technol. **57**(4), 2387–2394 (2008)
29. Xue, K., Li, S., Hong, J., Xue, Y., Yu, N., Hong, P.: Two-cloud secure database for numeric-related SQL range queries with privacy preserving. IEEE Trans. Inf. Forensics Secur. **12**(7), 1596–1608 (2017)
30. Zhang, Y., Yu, R., Xie, S., Yao, W., Xiao, Y., Guizani, M.: Home M2M networks: architectures, standards, and qos improvement. IEEE Commun. Mag. **49**(4), 44–52 (2011)
31. Xue, K., Xue, Y., Hong, J., Li, W., Yue, H., Wei, D.S.L., Hong, P.: RAAC: robust and auditable access control with multiple attribute authorities for public cloud storage. IEEE Trans. Inf. Forensics Secur. **12**(4), 953–967 (2017)
32. Zhang, Y., Yu, R., Nekovee, M., Liu, Y., Xie, S., Gjessing, S.: Cognitive machine-to-machine communications: visions and potentials for the smart grid. IEEE Netw **26**(3), 6–13 (2012)
33. Hei, X., Du, X.: Biometric-based two-level secure access control for implantable medical devices during emergencies. In: IEEE Conference on Computer Communications (INFO-COM), pp. 346–350 (2011)
34. Abdrashitov, A., Spivak, A.: Sensor data anonymization based on genetic algorithm clustering with L-diversity. In: 18th Conference of Open Innovations Association and Seminar on Information Security and Protection of Information Technology (FRUCT-ISPIT), pp. 3–8 (2016)
35. Shen, J., Zhou, T., Chen, X., Li, J., Susilo, W.: Anonymous and traceable group data sharing in cloud computing. IEEE Trans. Inf. Forensics Secur. **13**(4), 912–925 (2018)
36. Zhang, C., Zhu, L., Xu, C.: PTBI: an efficient privacy-preserving biometric identification based on perturbed term in the cloud. Inf. Sci. **409**, 56–67 (2017)
37. Du, X., Chen, H.-H.: Security in wireless sensor networks. IEEE Wirel. Commun. **15**(4), 60–66 (2008)
38. Du, X., Guizani, M., Xiao, Y., Chen, H.-H.: A routing-driven elliptic curve cryptography based key management scheme for heterogeneous sensor networks. IEEE Trans. Wirel. Commun. **8**(3), 1223–1229 (2009)

39. Li, J., Huang, X., Li, J., Chen, X., Xiang, Y.: Securely outsourcing attribute-based encryption with checkability. IEEE Trans. Parallel Distrib. Syst. **25**(8), 2201–2210 (2014)
40. Li, J., Li, J., Chen, X., Jia, C., Lou, W.: Identity-based encryption with outsourced revocation in cloud computing. IEEE Trans. Comput. **64**(2), 425–437 (2015)
41. Zhang, C., Zhu, L., Xu, C., Sharif, K., Du, X., Guizani, M.: LPTD: achieving lightweight and privacy-preserving truth discovery in CIoT. Futur. Gener. Comput. Syst. **90**, 175–184 (2019)
42. Liu, X., Qin, B., Deng, R.H., Lu, R., Ma, J.: A privacy-preserving outsourced functional computation framework across large-scale multiple encrypted domains. IEEE Trans. Comput. **65**(12), 3567–3579 (2016)
43. Bresson, E., Catalano, D., Pointcheval, D.: A simple public-key cryptosystem with a double trapdoor decryption mechanism and its applications. In: International Conference on the Theory and Application of Cryptology and Information Security, pp. 37–54 (2003)
44. Cheng, K., Shen, Y., Wang, Y., Wang, L., Ma, J., Jiang, X., Su, C.: Strongly secure and efficient range queries in cloud databases under multiple keys. In: IEEE Conference on Computer Communications (INFOCOM), pp. 2494–2502. IEEE, Piscataway (2019)
45. Dong, C., Wang, Y., Aldweesh, A., McCorry, P., van Moorsel, A.: Betrayal, distrust, and rationality: smart counter-collusion contracts for verifiable cloud computing. In: Proceedings of the 2017 ACM SIGSAC Conference on Computer and Communications Security, pp. 211–227 (2017)
46. Chen, C., Wang, X., Han, W., Zang, B.: A robust detection of the sybil attack in urban vanets. In: 29th IEEE International Conference on Distributed Computing Systems Workshops, pp. 270–276. IEEE, Piscataway (2009)
47. Guette, G., Ducourthial, B.: On the sybil attack detection in vanet. In: IEEE International Conference on Mobile Adhoc and Sensor Systems, pp. 1–6. IEEE, Piscataway (2007)
48. Hua, Y., Chen, F., Deng, S., Duan, S., Wang, L.: Secure distributed estimation against false data injection attack. Inf. Sci. **515**, 248–262 (2020)
49. James, J.Q., Hou, Y., Li, V.O.K.: Online false data injection attack detection with wavelet transform and deep neural networks. IEEE Trans. Ind. Inf. **14**(7), 3271–3280 (2018)
50. Rahman, Md.A., Mohsenian-Rad, H.: False data injection attacks with incomplete information against smart power grids. In: IEEE Global Communications Conference (GLOBECOM), pp. 3153–3158. IEEE, Piscataway (2012)
51. Lu, R., Heung, K., Lashkari, A.H., Ghorbani, A.A.: A lightweight privacy-preserving data aggregation scheme for fog computing-enhanced IoT. IEEE Access **5**, 3302–3312 (2017)
52. Goldreich, O.: Foundations of cryptography: volume 2, basic applications. Cambridge University Press, Cambridge (2009)
53. Chen, S., Li, M., Ren, K., Fu, X., Qiao, C.: Rise of the indoor crowd: reconstruction of building interior view via mobile crowdsourcing. In: Proceedings of the 13th ACM Conference on Embedded Networked Sensor Systems, pp. 59–71 (2015)

# Chapter 6
# Privacy-Preserving Truth Discovery with Truth Hiding

**Abstract** This chapter considers the participation of data requesters in mobile crowdsensing (MCS) data truth discovery scenarios and discusses how to achieve privacy-preserving truth discovery with truth hiding. In most MCS scenarios, the data requesters submit the sensing task to the mobile crowdsensing platform for publication, and the mobile crowdsensing platform collects the sensory data of sensing users. In this case, the data requesters want to obtain the task's truth, but the sensory data and the outputs of truth discovery may contain sensitive information and cause serious privacy concerns. Although existing privacy-preserving truth discovery schemes adopt several effective technologies to preserve the privacy of data and weights, most schemes do not consider the truth privacy. To this end, this chapter proposes a privacy-preserving truth discovery scheme SATE with truth hiding. Specifically, we build a mobile crowdsensing platform with two non-colluding servers, adopting the public-key cryptosystem supporting distributed decryption (PCDD) and using its homomorphic properties so users do not need to participate in the iterative process of privacy-preserving truth discovery. Security analysis proves that SATE effectively preserves the privacy of sensory data, weights, and truth. Performance evaluation and analysis prove that SATE has high computing and communication efficiency.

## 6.1 Introduction

This section describes the overview, related works, and preliminary of this chapter.

### 6.1.1 Overview

The rapid development of mobile crowdsensing (MCS) makes data collection more accessible and effortless. In a typical mobile crowdsensing application, the data requester uploads the sensing task to the mobile crowdsensing platform, and then the mobile crowdsensing platform collects the sensory data of sensing users. After data processing, the result is returned to the data requester. In order to get the task's true values (i.e., truth), the mobile crowdsensing platform usually applies the truth discovery algorithm [1–3] to process the sensory data. However, in running the truth discovery algorithm, the data and weight privacy of sensing users may be leaked. Although some researchers have proposed a series of privacy-preserving truth discovery schemes, most existing schemes do not consider the privacy-preserving of truth, which may compromise the privacy or economic rights and interests of data requesters. For example, Alice wants to know the traffic conditions of the destination area. Therefore, she issues traffic condition collection tasks for some locations to the mobile crowdsensing platform (the scheme in Chaps. 3 or 4 can be used to protect the task and location privacy). After receiving the data of sensing users, although the mobile crowdsensing platform can design a privacy-preserving truth discovery scheme as in Chap. 5 to get real traffic conditions at relevant locations since the truth is opened to the mobile crowdsensing platform, the mobile crowdsensing platform may still obtain the location or track privacy of the data requester based on the matching relationship between truth and background information. In addition, the data requesters usually need to pay for the sensing task, and the mobile crowdsensing platform may sell the truth to other companies or individuals for more benefits after obtaining the truth, which seriously damages the economic interests of data requesters. Therefore, when data requesters participate in the mobile crowdsensing application, preserving the sensory data and weights, strict preserving of the truth is also required.

However, implementing such a truth discovery scheme that can provide comprehensive privacy preservation faces many challenges from privacy and efficiency. Generally speaking, truth discovery includes weight update and truth update: (1) sensing users first estimate their weights by calculating the distance between their data and the truth, and (2) the mobile crowdsensing platform updates the truth by using a weighted average based on the sensing users' weight. But due to the unstable network status and strong mobility of sensing users, sensing users may not always be online. In order to avoid frequent interactions between sensing users and mobile crowdsensing platforms, current scholars are committed to implementing a privacy-preserving truth discovery scheme in which users do not need to participate in the iterative process of truth discovery [4–7]. The basic idea of these schemes is to use two no colluding cloud servers to interact with each other to update the truth and weights iteratively. However, most of these schemes do not consider the truth privacy of data requesters and have high computational costs and communication overheads.

This chapter, to address the above challenges, proposes a privacy-preserving truth discovery scheme based on two no colluding cloud servers (privacy-preserving non-interactive truth discovery, SATE) for preserving the truth, weights, and sensory data. Moreover, the user does not need to participate in the iterative truth discovery process. The contributions of this chapter are summarized as follows:

- For the scenario where data requesters participate in the mobile crowdsensing application, a privacy-preserving truth discovery scheme SATE is proposed, in which users do not need to participate in the iterative process of truth discovery. This scheme can provide comprehensive privacy preservation for the truth of the data requesters, the sensory data, and the weights of the sensing users.

- A lightweight data perturbation algorithm is designed to preserve the data privacy of sensing users. Compared with existing privacy-preserving truth discovery schemes, SATE achieves lower computational cost and communication overhead on the side of sensing users. In addition, based on a public-key cryptosystem (PCDD) on data perturbation technologies and supporting multi-party cooperative decryption, a privacy-preserving truth discovery algorithm is designed, which the cloud server can update the truth and weights in a privacy-preserving way without the participation of sensing users and data requesters.
- Extensive evaluations are conducted on real and simulated mobile crowdsensing scenarios. The analysis results prove that SATE has low computational cost and communication overhead on both the user and the mobile crowdsensing platform sides.

The rest of this chapter is organized as follows: Sect. 6.1.2 reviews the related works on privacy-preserving truth discovery schemes based on two non-collusion cloud servers; Sect. 6.1.3 provides a brief description of preliminaries; Sect. 6.2 describes the architecture overview of SATE from system model, security model, and design goals; Sect. 6.3 presents the detailed design of SATE; Sect. 6.4 gives the security analysis of SATE; Sect. 6.5 presents the performance evaluation and analysis; and Sect. 6.6 summarizes the work of this chapter.

## *6.1.2 Related Works*

In order to reduce the computational cost and communication overhead of sensing users, researchers have proposed multiple privacy-preserving truth discovery schemes based on two non-collusion cloud servers [4–7]. Specifically, Miao et al. [4] proposed the first privacy-preserving truth discovery scheme $L^2$-PPTD in which users do not need to participate in the iterative process of truth discovery. Although $L^2$-PPTD can significantly reduce users' computational cost and communication overhead, it ignores the preservation of user weights. To realize the dual preservation of user sensory data and weights, Tang et al. [5], Tang et al. [6], and Zhang et al. [7] based on garbled circuits, Paillier homomorphic encryption algorithm, data perturbation, and other technologies, respectively, proposed an effective solution. However, none of these schemes consider the privacy preservation of the truth. In addition, in order to transfer the computation in the iterative process of truth discovery to the mobile crowdsensing platform, most of the above schemes require sensing users to perform complex cryptographic operations in the initialization phase, which increases the computational cost and communication overhead of the sensing user. Table 6.1 lists the comparison of SATE and existing schemes in terms of privacy and efficiency.

**Table 6.1** Comparison of privacy and efficiency properties between SATE and existing schemes

|                                  | Scheme [4] | Scheme [5] | Scheme [6] | Scheme [7] | SATE |
|----------------------------------|------------|------------|------------|------------|------|
| Sensory data privacy-preserving  | ✓          | ✓          | ✓          | ✓          | ✓    |
| Weight privacy-preserving        | ✗          | ✓          | ✓          | ✓          | ✓    |
| Truth privacy-preserving         | ✗          | ✗          | ✗          | ✗          | ✓    |
| Sensing user efficiency          | High       | Low        | High       | Low        | High |

## 6.1.3 Preliminary

### 6.1.3.1 Polynomial Function

Given a polynomial function $f(x) = a_n x^n + a_{n-1} x^{n-1} + \cdots + a_1 x + a_0$ with the highest power of $n$, where the coefficient of $x^i$ is $a_i$, $i \in [0, n]$, and $a_n \neq 0$, according to the basic theorem of algebra,

$$f(x) = a_n x^n + a_{n-1} x^{n-1} + \cdots + a_1 x + a_0$$
$$= a_n (x - x_n)(x - x_{n-1}) \cdots (x - x_1), \tag{6.1}$$

where $(x_1, x_2, \cdots, x_n)$ is the root of polynomial function $f(x)$. According to Weida's theorem, the $n - k$th coefficient of $f(x)$, i.e., $a_{n-k}(k \in [0, n])$, can be calculated as

$$\sum_{1 \leq i_1 < i_2 < \cdots < i_k \leq n} \left( \prod_{j=1}^{k} x_{i_j} \right) = (-1)^k \frac{a_{n-k}}{a_n}. \tag{6.2}$$

In the privacy-preserving task allocation scheme based on task contents designed in this book, a polynomial function is used to judge whether a task $x_i$ belongs to the task set $\{x_1, x_2, \cdots, x_n\}$. That is, suppose $\{\texttt{Task}(m) \to x_m\}_{m=1}^{n}$ is the required task set, and $f(x) = a_n (x - x_n)(x - x_{n-1}) \cdots (x - x_1)$ is the corresponding polynomial function. If $f(x_t) = 0$, the task $\texttt{Task}(m) \to x_t$ is in the required task set $\{\texttt{Task}(m) \to x_m\}_{m=1}^{n}$.

### 6.1.3.2 Searchable Encryption

Searchable encryption (SE) enables the third-party platform to find data or users that meet the requirements of data requesters in the form of ciphertext. From the perspective of encryption algorithms used, searchable encryption can be roughly divided into symmetric searchable encryption and public-key searchable encryption. This part mainly introduces the framework of searchable encryption scheme. On the whole, the searchable encryption scheme includes three entities: third-party platform, data holder, and data requester. Its basic process is summarized as follows:

- System Initialization: the data holder generates the key $sk$ and shares it with the data requester.
- Data Encryption: the data holder encrypts the data $w$ with the key and sends the ciphertext $E[w]$ to the third-party platform.
- Trapdoor Generation: for the search keyword $w'$, the data requester uses the key to generate $T[w']$ and send it to the third-party platform.
- Data Retrieval: the third-party platform calculates the ciphertext $E[w]$ and trapdoor $T[w']$. If $w = w'$, output 1; otherwise, output 0.

If the SE scheme outputs 1 for any $w = w'$, then the SE scheme is correct. This book uses the SE framework to a design privacy-preserving task allocation scheme.

### 6.1.3.3 Truth Discovery

A truth discovery algorithm is an effective technology to calculate the real results of each task from multi-task data. On the whole, truth discovery mainly includes two steps: weight update and truth update. Suppose there are $K$ users and $M$ perception tasks, the perception data of user $u_k$ for task $o_m$ is $x^k_m$, the weight of user $u_k$ is $w_k$, and the truth of task $o_m$ is $x^*_m$. The following two steps are iterated until the iteration ends, and the truth of each task can be obtained:

- Weight Update: this step calculates the weight of the user according to the user data and the task truth. Specifically, the user's weight can be calculated as

$$w_k = f\left(\sum_{m=1}^{M} d\left(x^k_m, x^*_m\right)\right), \tag{6.3}$$

where $f(\cdot)$ is a monotonically decreasing function, and $d(x^k_m, x^*_m)$ represents the distance between user data and truth.
- Truth Update: this step calculates the truth of the task according to the user's weight and data. Specifically, the truth can be calculated as

$$x^*_m = \frac{\sum_{k=1}^{K} w_k \cdot x^k_m}{\sum_{k=1}^{K} w_k}. \tag{6.4}$$

In weight updating, the distance calculation function $d(\cdot)$ is slightly different according to different data types. For continuous data (such as speed, temperature, humidity, etc.), the square distance function $d(x^k_m, x^*_m) = (x^k_m - x^*_m)^2$ is used to represent the distance between user data and truth; for discrete data (such as question and answer options), assume that task $o_m$ has several answer options, $x^k_m = (0, \cdots, \frac{1}{q}, \cdots, 0)^T$ represents the q-th answer selected by user $u_k$, and

$d(x^k_m, x^*_m) = (x^k_m - x^*_m)^T (x^k_m - x^*_m)$ represents the distance between user data and truth. Consistent with most of the existing truth discovery schemes [8, 9], this book selects the logarithmic equation to calculate the user's weight, i.e.,

$$w_k = \log \left( \frac{\sum_{k=1}^{K} \sum_{m=1}^{M} d\left(x^k_m, x^*_m\right)}{\sum_{m=1}^{M} d\left(x^k_m, x^*_m\right)} \right). \tag{6.5}$$

In the privacy protection truth discovery scheme of truth disclosure and truth hiding designed in this book, the truth discovery algorithm is used to calculate the truth of the task.

### 6.1.3.4 Public-Key Cryptosystem Supporting Distributed Decryption

Liu et al. [10] proposed the public-key cryptosystem supporting distributed decryption (PCDD) based on the public-key encryption algorithm [11]. PCDD splits the decrypted private key and distributes it to several entities. Multiple entities decrypt ciphertext through cooperation. Because of its high efficiency and scalability, PCDD is widely used in secure multi-party computing and other applications. It mainly includes the following algorithms:

- **KeyGen**($\kappa$)(Initialization): given a security parameter $\kappa$, select two large prime numbers $p, q, |p| = |q| = \kappa$ and calculate $n = pq$, $\lambda = lcm(p-1)(q-1)/2$, where $lcm(a, b)$ is the least common multiple of $a$ and $b$. Select random number $g \in Z^*_{n^2}$ and random number $\theta \in [1, n/4]$, and calculate $h = g^\theta \bmod n^2$. The public key is $pk = (n, g, h)$, and the weak private key is $\theta$. The strong private key is $\lambda$.
- **Enc**($pk, m$)(Data Encryption): given the data $m \in Z_n$, the algorithm encrypts $m$ as $\mathbf{E}[m] = h^r(1 + nm) \bmod n^2, r \in [1, n/4]$.
- **SDec**($\lambda, \mathbf{E}[m]$)(Ciphertext Decryption): given the ciphertext $\mathbf{E}[m]$, the algorithm uses a strong private key to decrypt the ciphertext $m = L((\mathbf{E}[m])^\lambda \bmod n^2)\lambda^{-1} \bmod n$, where $(\mathbf{E}[m])^\lambda \bmod n^2 = g^{\theta r \lambda}(1+mn\lambda) \bmod n^2$, $L(x) = \frac{x-1}{n}$.
- **SkeyS**($\lambda, t$)(Private Key Decomposition): given the strong private key $\lambda$, the algorithm divides $\lambda$ into t parts $(\lambda_1, \lambda_2, \cdots, \lambda_t)$, where $\sum_{i=1}^{t} \lambda_i \equiv 0 \bmod \lambda$, $\sum_{i=1}^{t} \lambda_i \equiv 1 \bmod n^2$.
- **PSDec**($\lambda, \mathbf{E}[m]$)(Partial Decryption): given the ciphertext $\mathbf{E}[m]$, the algorithm partially decrypts the ciphertext as $CT^{(i)} = (\mathbf{E}[m])^\lambda_i = g^{\theta r \lambda_i}(1+mn\lambda_i) \bmod n^2$.
- **DDec**($\{CT^{(i)}\}_{i=1}^{t}$)(Complete Decryption): given the partially decrypted ciphertext $CT^{(1)}, CT^{(2)}, \cdots, CT^{(t)}$, the algorithm performs complete decryption $m = L(\prod_{i=1}^{t} CT^{(i)})$.

For any $m \in Z_n$, $(\mathbf{E}[m])^{n-1} = h^{r(n-1)} \cdot (1 + (n-1)mn) \bmod n^2 = \mathbf{E}[-m]$. For the convenience of narration, this book uses $\mathbf{E}[m])^{-1}$ to represent $\mathbf{E}[m])^{n-1}$. PCDD has a homomorphic property similar to Paillier encryption system, that is, given plaintext $m_1, m_2, a \in Z_n$, PCDD satisfies

$$E[m_1] \cdot E[m_2] = E[m_1 + m_2]$$
$$(E[m])^a = E[a \cdot m]. \tag{6.6}$$

PCDD is used to protect data privacy and construct privacy protection truth discovery algorithm.

## 6.2   Architecture Overview

This section describes the system model, security model, and design goals of SATE.

### 6.2.1   System Model

As shown in Fig. 6.1, the system model of SATE mainly includes three entities: data requester, mobile crowdsensing platform, and sensing user. The roles and functions of each entity are briefly described as follows:

- Data requester: The data requester can be an individual, a company, or an organization. They want to get truthful sensory data for several sensing tasks. However, they cannot be obtained by themselves due to location, ability, and

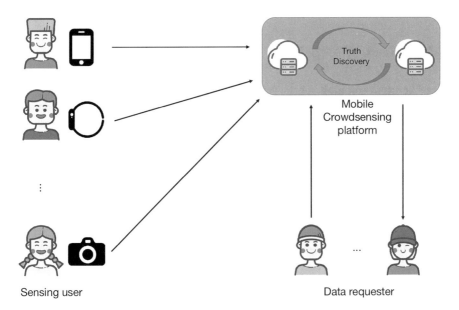

**Fig. 6.1**  System model of SATE

background knowledge limitations. Thus, the data requester outsources the sensing tasks to a mobile crowdsensing platform. The platform helps them publish tasks and collect data. Since the data requester usually pays for the sensing tasks, the truth cannot be disclosed to other entities.

- Mobile crowdsensing platform: The mobile crowdsensing platforms are assumed to have enormous data collection, calculation, and transmission resources. It is responsible for connecting data requesters and sensing users. To ensure the quality of service, when the sensory data is collected from the sensing user, the mobile crowdsensing platforms should filter out real results from existing datasets and return them to data requesters. In SATE, we use two no colluding cloud servers (such as Amazon Website Service and Microsoft Azure) as the mobile crowdsensing platform.
- Sensing user: The sensing users, as data providers, accept tasks from the mobile crowdsensing platform and use sensors to collect data. After the data collection, the sensing user uploads the data to the mobile crowdsensing platform. Since sensing devices usually have limited computing and communication resources, the overhead of sensing users should be as small as possible.

The problem we target in this chapter can be summarized supposing there are $M$ sensing tasks published by a data requester and $K$ sensing users involved in the system. Let $x_m^k$ represent the sensory data submitted by a sensing user $u_k$ for the task $o_m$, $w_k$ represent the user's weight, and $x_m^*$ represent the truth of $o_m$. SATE's goal is to let two clouds, $S_A$ and $S_B$, cooperatively calculate the truth based on the data they receive from users in a privacy-preserving way. Sensing users' sensory data, weight, and estimated truths should be strictly protected during this process.

## 6.2.2   Security Model

Sensing users and mobile crowdsensing platforms participating in the system are considered semi-trusted. All entities will perform the designed protocol honestly but be curious about participants' privacy, including sensory data, weights, and truths. Moreover, there is no collusion between two cloud servers, while anyone may collude with several sensing users. This assumption is consistent with some existing schemes based on the two-server model [5, 7]. It is reasonable since there are often conflicts of interest between cloud service providers. In addition, we can also employ mechanisms such as game theory [12] to reduce the possibility of collusion between two cloud service providers. Since the data requester is the consumer of data and needs to pay for sensory data and truth-seeking services, we suppose that the data requester is trusted, and the data requester will not actively acquire the data of the sensing user. In addition, this chapter does not consider maliciously providing multiple data or fake data injection attacks. The former is similar to the Sybil attack and can be solved by scheme [13], while the latter can be solved by authenticating the identity of the sensing user [14].

## 6.2.3 Design Goals

SATE is designed to balance privacy-preserving and efficiency. Specifically, SATE aims to achieve the following goals:

- Privacy: SATE should ensure that the truth of the data requester and the sensory data and weight of the sensing user are effectively preserved so that the mobile crowdsensing platform cannot obtain the privacy of the sensing user and the data requester based on the existing data.
- Efficiency: SATE should minimize the computational cost and communication overhead on the sensing user and mobile crowdsensing platform sides. In particular, SATE should ensure that sensing users and data requesters do not need to be involved in the iterative process of truth discovery.

## 6.3 Detailed Design

This section will introduce the detailed design of SATE. Specifically, SATE mainly includes two parts: initialization and iteration. For the convenience of description, we use PCDD to represent the public-key cryptosystem supporting distributed decryption, $E[x]$ to denote the ciphertext obtained by encrypting $x$ using the PCDD encryption algorithm, $D_s k[x]$ to denote the information obtained after decrypting $x$ with the private key $sk$ using the PCDD decryption algorithm, $PSDec(sk, C)$ to denote the partial decryption operation on the ciphertext $C$ using the private key $sk$ using the PCDD partial decryption algorithm, and $DDSec(C1, C2)$ to denote that the ciphertext $C_1, C_2$ is completely decrypted using the PCDD complete decryption algorithm. The specific description of the above-related algorithms has been given in Chap. 2 and will not be repeated in this chapter. We use $S_A, S_B$ to denote two cloud servers respectively.

### 6.3.1 Initialization

**Step 1**

Given a security parameter $\kappa$, the data requester $u_i$ runs $PCDD.KeyGen(\kappa)$ to get the public key $(n, g, h)$ and private key $\lambda$, and runs $PCDD.SkeyS(\lambda, 2)$ to split the private key into two parts $(\lambda_A, \lambda_B)$. Subsequently, $u_i$ discloses the public key and sends $\lambda_A, \lambda_B$ to $S_A, S_B$ cloud servers respectively through a secure channel.

**Step 2**

$u_i$ generates random truths $[x_1^*; \ldots; x_M^*]$, the random number $\{\beta_m\}_{m=1}^M$ is selected, the perturbed truth is $\tilde{x}_m^* = x_m^* + \beta_m$, and the encrypted random number is $E[\beta_m], E[\beta_m^2]$. Then, $u_i$ sends the perturbed truth $\{\tilde{x}_m^*\}_{m=1}^M$ to $S_A$ and sends the ciphertext $E[\beta_m], E[\beta_m^2]_{m=1}^M$ to $S_B$.

**Step 3**

The sensing user $u_k$ generates a series of random numbers $\{\alpha_m^k\}_{m=1}^M$ for the collected sensory data $[x_1^k, \ldots, x_M^k]$ and perturbs it to obtain $\tilde{x}_m^k = x_m^k - \alpha_m^k$. Then, $u_k$ sends the perturbed value $\{\tilde{x}_m^k\}_{m=1}^M$ to $S_A$ and the random number $\{\alpha_m^k\}_{m=1}^M$ to $S_B$.

**Step 4**

After receiving the data of the sensing user, $S_B$ computes $E[\alpha_m^k] \cdot E[\beta_m]$ to obtain the ciphertext of $\alpha_m^k + \beta_m$ and computes $E[\sum_{m=1}^M (\alpha_m^k)^2] \cdot \prod_{m=1}^M (E[\beta_m]^{2\alpha_m^k} \cdot E[\beta_m^2])$ to obtain the ciphertext of $\sum_{m=1}^M (\alpha_m^k + \beta_m)^2$. Subsequently, $S_B$ sends the ciphertext $\{E[\alpha_m^k + \beta_m]\}_{k,m=1}^{K,M}$ and the ciphertext $\{E[\sum_{m=1}^M (\alpha_m^k + \beta_m)^2]\}_{k=1}^K$ to $S_A$.

The above steps are performed only once in SATE. Note that the sensory data or random number that needs to be encrypted may not be an integer. In this case, we can use the coefficient $L$ ($L$ is usually a multiple of 10) to round up the decimal and, after getting the final result, divide it to get the correct result. Figure 6.2 shows the detailed steps in the initialization phase of the SATE system.

### *6.3.2  Iteration*
**Step 1**

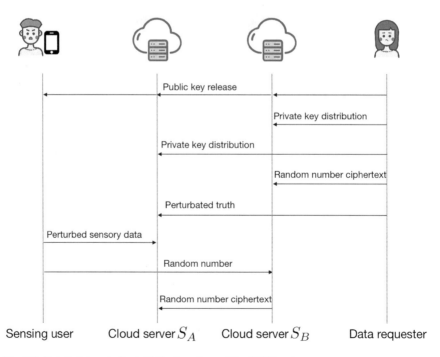

**Fig. 6.2** Detailed steps in the initialization phase of the SATE system

Based on the data obtained during the initialization phase, $S_A$ computes

$$C_d^k = \mathsf{E}\left[\sum_{m=1}^{M}(\tilde{x}_m^k - \tilde{x}_m^*)^2\right] \cdot \mathsf{E}\left[\sum_{m=1}^{M}(\alpha_m^k + \beta_m)^2\right] \cdot \prod_{m=1}^{M}\mathsf{E}\left[\alpha_m^k + \beta_m\right]^{2(\tilde{x}_m^k - \tilde{x}_m^*)}$$

$$= \mathsf{E}\left[\sum_{m=1}^{M}(x_m^k - x_m^* - (\alpha_m^k + \beta_m)^2 + (\alpha_m^k + \beta_m)^2)\right.$$

$$\left. + (\alpha_m^k + \beta_m)\cdot(2(x_m^k - x_m^* - (\alpha_m^k + \beta_m)))\right]$$

$$= \mathsf{E}\left[\sum_{m=1}^{M}(x_m^k - x_m^*)^2\right]. \tag{6.7}$$

After obtaining the distance ciphertext between user $uk$'s sensory data $\{x_m^k\}_{m=1}^M$ and the truth $\{x_m^*\}_{m=1}^M$, then $S_A$ aggregates the distance ciphertexts of all users to get

$$C_D = \prod_{k=1}^{K}C_d^k$$

$$= \mathsf{E}[\sum_{k=1}^{K}\sum_{m=1}^{M}(x_m^k - x_m^*)^2], \tag{6.8}$$

and send it to $S_B$.

**Step 2**
According to private key $\lambda_B$, $S_B$ performs partial decryption on the received ciphertext.

$$\mathsf{D}_B[C_D] \leftarrow \mathsf{PSDec}(\lambda_B, C_D). \tag{6.9}$$

Then, $S_B$ returns $\mathsf{D}_B[C_D]$ to $S_A$.

**Step 3**
After receiving the partially decrypted ciphertext, $S_A$ use its private key $\lambda_A$ to compute

$$\mathsf{D}_A[C_D] \leftarrow \mathsf{PSDec}(\lambda_A, C_D) \tag{6.10}$$

and restore the aggregated distance of all users as

$$\sum_{k=1}^{K}\sum_{m=1}^{M}(x_m^k - x_m^*)^2 \leftarrow \mathsf{DDSec}(\mathsf{D}_A[C_D], \mathsf{D}_B[C_D]). \tag{6.11}$$

For each sensing user $u_k$, $S_A$ chooses a random number $r_A^k \in Zn$ to compute

$$\tilde{C}_d^k = (C_d^k)^{\frac{r_A^k}{\sum_{k=1}^{K}\sum_{m=1}^{M}(x_m^k - x_m^*)^2}}$$
$$= \mathsf{E}\left[\frac{r_A^k \sum_{m=1}^{M}(x_m^k - x_m^*)^2}{\sum_{k=1}^{K}\sum_{m=1}^{M}(x_m^k - x_m^*)^2}\right]. \tag{6.12}$$

Subsequently, $S_A$ partially decrypts $\tilde{C}_d^k$

$$\mathsf{D}_A[\tilde{C}_d^k] \leftarrow \mathsf{PSDes}(\lambda_A, \tilde{C}_d^k) \tag{6.13}$$

and sends $\{\tilde{C}_d^k, \mathsf{D}_A[\tilde{C}_d^k]\}_{k=1}^{K}$ to $S_B$.

**Step 4**
$S_B$ uses its private key $\lambda_B$ to partially decrypt the ciphertext $\tilde{C}_d^k$

$$\mathsf{D}_B[\tilde{C}_d^k] \leftarrow \mathsf{PSDes}(\lambda_B, \tilde{C}_d^k) \tag{6.14}$$

and completely decrypt the ciphertext $\tilde{C}_d^k$ according to $\mathsf{D}_A[\tilde{C}_d^k]$

$$\frac{r_A^k \sum_{m=1}^{M}(x_m^k - x_m^*)^2}{\sum_{k=1}^{K}\sum_{m=1}^{M}(x_m^k - x_m^*)^2} \leftarrow \mathsf{DDSec}(\mathsf{D}_A[\tilde{C}_d^k], \mathsf{D}_B[\tilde{C}_d^k]). \tag{6.15}$$

Based on $\frac{r_A^k \sum_{m=1}^{M}(x_m^k - x_m^*)^2}{\sum_{k=1}^{K}\sum_{m=1}^{M}(x_m^k - x_m^*)^2}$, the weight of sensing user $u_k$ can be updated to

$$\overline{w}_k = -log\left(\frac{r_A^k \sum_{m=1}^{M}(x_m^k - x_m^*)^2}{\sum_{k=1}^{K}\sum_{m=1}^{M}(x_m^k - x_m^*)^2}\right) + r_B^k$$
$$= log\left(\frac{\sum_{k=1}^{K}\sum_{m=1}^{M}(x_m^k - x_m^*)^2}{r_A^k \sum_{m=1}^{M}(x_m^k - x_m^*)^2}\right) + r_B^k \tag{6.16}$$
$$= w_k - log(r_A^k) + r_B^k,$$

where $r_B^k \in Z_n$ is a random number used to perturb the weights. Subsequently, $S_B$ computes

$$\overline{C}_m = \prod_{k=1}^{K} \mathsf{E}[\alpha_m^k + \beta_m]^{-r_B^k} \qquad (6.17)$$

and returns the ciphertext $(\{\overline{C}_m\}_{m=1}^{M}, \{\mathsf{E}[r_B^k]\}_{k=1}^{K})$ and the perturbed weight $\{\overline{w}_k\}_{k=1}^{K}$ to $S_A$.

**Step 5**
According to $r_A^k$, $S_A$ computes

$$\tilde{w}_k = \overline{w}_k + log(r_A^k)$$
$$= w_k - log(r_A^k) + r_B^k + log(r_A^k) \quad = w_k + r_B^k \qquad (6.18)$$

and partially restores the weight $w_k$ of sensing user $u_k$. Subsequently, $S_A$ computes the sum of all perturbation values and

$$\tilde{C}_m = \left( \prod_{k=1}^{K} \mathsf{E}[\alpha_m^k + \beta_m]^{\tilde{w}_k} \right) \cdot \overline{C}_m \cdot \prod_{k=1}^{K} \mathsf{E}[r_B^k]^{-\tilde{x}_m^k}$$

$$= \left( \prod_{k=1}^{K} \mathsf{E}[\alpha_m^k + \beta_m]^{\tilde{w}_k} \right) \cdot \prod_{k=1}^{K} \mathsf{E}[\alpha_m^k + \beta_m]^{-r_B^k} \cdot \prod_{k=1}^{K} \mathsf{E}[r_B^k]^{-\tilde{x}_m^k}$$

$$= \prod_{k=1}^{K} \mathsf{E}[\alpha_m^k + \beta_m]^{\tilde{w}_k} \cdot \mathsf{E}[\alpha_m^k + \beta_m]^{-r_B^k} \cdot \mathsf{E}[r_B^k]^{-\tilde{x}_m^k} \qquad (6.19)$$

$$= \mathsf{E}\left[ \sum_{k=1}^{K} ((\tilde{w}_k - r_B^k)(\alpha_m^k + \beta_m) - r_B^k \tilde{x}_m^k) \right]$$

$$= \mathsf{E}\left[ \sum_{k=1}^{K} ((w_k(\alpha_m^k + \beta_m)) - r_B^k \tilde{x}_m^k \tilde{x}_m^k) \right],$$

partially decrypts

$$\mathsf{D}_A[\tilde{C}_m] \leftarrow \mathsf{PSDec}(\lambda_A, \tilde{C}_m), \qquad (6.20)$$

and sends $\{\tilde{C}_m, \mathsf{D}_A[\tilde{C}]\}_{m=1}^{M}$ to $S_B$.

**Step 6**
After receiving the ciphertext, $S_B$ uses private key $\lambda_B$ to partially decrypt $\tilde{C}_m$

$$D_B[\tilde{C}_m] \leftarrow \mathsf{PSDec}(\lambda_B, \tilde{C}_m), \tag{6.21}$$

and completely decrypts according to $D_A[\tilde{C}_m]$

$$\tilde{r\alpha}_m \leftarrow \mathsf{DDSec}(D_A[\tilde{C}_m], D_B[\tilde{C}_m]). \tag{6.22}$$

Then, $S_B$ sends $(\{\tilde{r\alpha}_m\}_{m=1}^M, \tilde{rw} = \sum_{k=1}^K r_B^k)$ to $S_A$.

**Step 7**
Based on $\tilde{r\alpha}_m, \tilde{rw}$, $S_A$ updates the truth

$$
\begin{aligned}
\tilde{x}_m^* &= \frac{\sum_{k=1}^K \tilde{w}_k \cdot \tilde{x}_m^k + \tilde{r\alpha}_m}{\sum_{k=1}^K \tilde{w}_k - \tilde{rw}} \\
&= \frac{\sum_{k=1}^K (w_k + r_B^k)(x_m^k - \alpha_m^k) + \sum_{k=1}^K (w_k(\alpha_m^k + \beta_m) - r_B^k(x_m^k - \alpha_m^k))}{\sum_{k=1}^K (w_k + r_B^k) - \sum_{k=1}^K r_B^k} \\
&= \frac{\sum_{k=1}^K w_k \cdot (x_m^k + \beta_m)}{\sum_{k=1}^K w_k} \\
&= x_m^* + \beta_m.
\end{aligned}
\tag{6.23}
$$

$S_A$ and $S_B$ iteratively run **Steps 1–7**; when the iteration is terminated (such as the difference between two consecutive truths is less than a supposed threshold or reaches a predetermined number of iterations), $S_A$ sends the perturbation truth $\{\tilde{x}_m^*\}_{m=1}^M$ to the data requester $u_i$. $u_i$ can restore the truth through $x_m^* = \tilde{x}_m^* - \beta_m$. Figure 6.3 shows the detailed steps of SATE.

## 6.4 Security Analysis

This section presents the security analysis of SATE. Before starting a formal analysis, we first review the formal security definition of the semi-trusted adversary model given in Chap. 5 Definition 5.1. Specifically, we suppose that protocol $P$ requires $\mathcal{A}$ to compute $f_{\mathcal{A}}(x, y)$ and $\mathcal{B}$ to compute $f_{\mathcal{B}}(x, y)$, where $x, y$ are respectively the inputs to $\mathcal{A}$, $\mathcal{B}$. We suppose $view_{\mathcal{A}}(x, y), view_{\mathcal{B}}(x, y)$ to be the results observed by $\mathcal{A}$ and $\mathcal{B}$ for $(x, y)$ through protocol $P$. That is, if $(x, r_A)$ (or

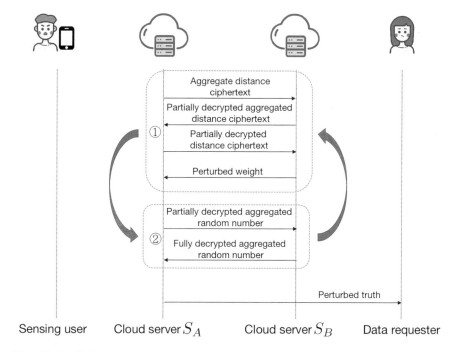

**Fig. 6.3** Detailed steps in the iteration phase of the SATE system

$(y, r_B))$ is the input and random number of $\mathcal{A}$ (or $\mathcal{B}$), $m_i$ is the $i^{th}$ message uploaded to protocol $P$:

$$view_{\mathcal{A}}(x, y) = (x, r_{\mathcal{A}}, m_1, m_2, \ldots m_t),$$
$$view_{\mathcal{B}}(x, y) = (x, r_{\mathcal{B}}, m_1, m_2, \ldots m_t). \tag{6.24}$$

Let $O_{\mathcal{A}}(x, y), O_{\mathcal{B}}(x, y)$ be the output of $\mathcal{A}, \mathcal{B}$. If there are two polynomial time (P.P.T) simulators $S_1, S_2$ making

$$(S_1(x, f_{\mathcal{A}}(x, y)), f_{\mathcal{B}}(x, y)) \equiv (view_{\mathcal{A}}(x, y), O_{\mathcal{B}}(x, y)),$$
$$(f_{\mathcal{A}}(x, y), S_2(y, f_{\mathcal{B}}(x, y))) \equiv (O_{\mathcal{A}}(x, y), view_{\mathcal{B}}(x, y)), \tag{6.25}$$

where $\equiv$ denotes that the computation is indistinguishable, then protocol $P$ can resist semi-trusted adversaries. It should be pointed out that the public-key cryptosystem PCDD used by SATE that supports multi-party cooperative decryption is already proved in scheme [10] to be semantically secure in a semi-trusted adversary model (i.e., ciphertext is indistinguishable). Based on the above definitions and conclusions, we give the following analysis.

**Theorem 6.1** *Suppose the encryption algorithm PCDD is semantically secure in the SATE scheme. In that case, the mobile crowdsensing platform cannot know the truth of the data requester and the sensory data and weight of the sensing user through the existing data.*

***Proof*** Since sensing users and data requesters do not need to participate in the iterative truth discovery process, we next prove that the mobile crowdsensing platform cannot restore truth, weight, and sensory data privacy through existing data.

For the cloud server $S_A$, the information it knows is ciphertext ($\mathsf{E}[\alpha_m^k + \beta_m]$, $\mathsf{E}[\sum_{m=1}^{M}(\alpha_m^k + \beta_m)^2]$, $\overline{C}_m$, $\mathsf{E}[r_B^k]$) and plaintext ($\tilde{x}_m^*$, $\tilde{x}_m^k$, $\sum_{k=1}^{K}\sum_{m=1}^{M}(x_m^k - x_m^*)^2$, $\tilde{w}_k$, $\tilde{r}\alpha_m$, $\tilde{r}\tilde{w}$). $S_A$ cannot decrypt the above ciphertext while holding only part of the private key. Since the random numbers $\beta_m$ and $\alpha_m^k$ are respectively generated by the data requester and the sensing user, without knowledge of the associated random numbers, $S_A$ cannot recover truth and sensing data from $\tilde{x}_m^*$ and $\tilde{x}_m^k$. Since $S_A$ can only collude with a small number of sensing users and $S_A$ does not know the truth $x_m^*$, $S_A$ also cannot obtain sensing data from the aggregated result $\sum_{k=1}^{K}\sum_{m=1}^{M}(x_m^k - x_m^*)^2$ either. Since $r_B^k$ is a random number generated by $S_B$, $S_A$ cannot restore any privacy from $\tilde{r}\alpha_m$ and $\tilde{r}\tilde{w}$. In addition, without knowing $r_B^k$, $S_A$ cannot restore the user's real weight information from the perturbed weight $\tilde{w}_k$.

For the cloud server $S_B$, the information it knows is ciphertext ($\mathsf{E}[\beta_m]$, $\mathsf{E}[(\beta_m)^2]$, $C_D$), and plaintext ($\frac{r_A^k \sum_{m=1}^{M}(x_m^k-x_m^*)^2}{\sum_{k=1}^{K}\sum_{m=1}^{M}(x_m^k-x_m^*)^2}$, $\overline{w}_k$). Without knowing the other private key $\lambda_A$, $S_B$ cannot restore the plaintext information from the ciphertext. Due to the unknown of perturbed sensing data $\tilde{x}_m^k$, $S_B$ cannot derive true sensory data from a known random number $\alpha_m^k$. Since $r_A^k$, $x_m^*$, $\sum_{k=1}^{K}\sum_{m=1}^{M}(x_m^k - x_m^*)^2$ are unknown, even though it may collude with some sensing users, $S_B$ cannot infer the sensory data of a single sensing user from $\frac{r_A^k \sum_{m=1}^{M}(x_m^k-x_m^*)^2}{\sum_{k=1}^{K}\sum_{m=1}^{M}(x_m^k-x_m^*)^2}$. Furthermore, since $S_B$ does not know $\log(r_A^k)$, it cannot know the sensing user's weight from $\overline{w}_K$. Therefore, it can be seen that SATE can effectively preserve the truth of the data requester as well as the sensing data and weight privacy of the sensing user.

The proof is over.

## 6.5  Performance Evaluation and Analysis

Specifically, we use a laptop with a 2.5 GHz Intel i5 processor and 16.0 GB RAM as the cloud server and an Android phone with 6.0 GB RAM as the sensing user and data requester. The experiments on the mobile phone side are implemented in Java, and the experiments on the laptop side are implemented in Python. The security parameter is set to 512 bits, and the encryption algorithm uses the public-key cryptosystem PCDD [10] that supports multi-party cooperative decryption. As a comparison, we also implement schemes EPTD [5] and RPTD-II [7] to compare

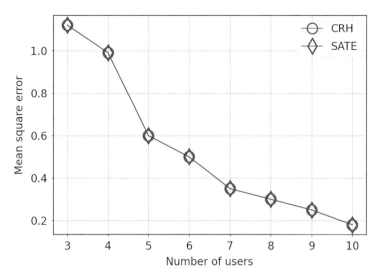

**Fig. 6.4** SATE correctness analysis

the computational and communication cost of SATE. It should be noted that EPTD is based on a dual-server design, while RPTD-II is based on the fog-cloud server design so we will implement these two schemes according to the design ideas of the scheme itself. Before giving specific experimental results, we first analyze the correctness and convergence of SATE.

- **Correctness analysis.** We first analyze the accuracy of SATE through the truth discovery algorithm CRH [8] without privacy-preserving. Similar to the scheme [4, 7], we choose the root mean squared error (RMSE) between the truth and the computed truth as the evaluation criterion. Specifically, in the simulation experiment, the number of observation tasks is set to 10, and the number of users is from 3 to 10. The performance of SATE is similar to that of CRH. It can be seen from Fig. 6.4 that the encryption operation of SATE does not sacrifice the functionality of truth discovery.

- **Convergence analysis.** Then, we analyze the convergence of SATE. The criterion of convergence is the difference between the truth of two consecutive updates $|x^t - x^{t-1}|$, where $x^t$ represents the truth of the $t$th iteration and $x^0$ is randomly generated. We randomly select several truths and perturb them by generating Gaussian noise based on the truths, and the threshold is set as 0.001. It can be seen from Fig. 6.5 that when the number of iterations is 10, the scheme can reach convergence.

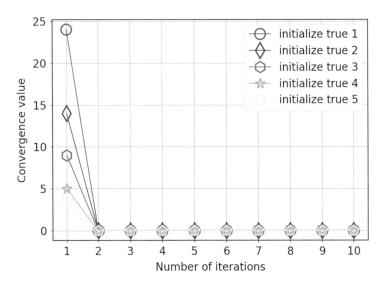

**Fig. 6.5** SATE convergence analysis

## 6.5.1  Computational Cost

### 6.5.1.1  Experiments Based on Real Mobile Crowdsensing Scenarios

We first design experiments on continuous data of real mobile crowdsensing scenarios. Specifically, we refer to scheme [13], which uses mobile phone sensors (such as gyroscope and compass) to collect the distances between different locations and upload them to the mobile crowdsensing platform. In this experiment, we use 10 mobile phones as sensing users and select the distance of 20 corridors as the observation task. Each experiment was run 10 times, and the average value was selected as the experimental result.

Figures 6.6 and 6.7 show the computational cost required for sensing users' and cloud servers' sides. Figure 6.6 shows that with the increase in the number of encryption tasks, the computational cost of the three schemes on the sensing user side increases. Since SATE uses data perturbation technology to preserve data privacy, it can significantly save the computational cost of mobile terminals compared with schemes based on Paillier homomorphic encryption [6, 7]. For example, when the number of encryption tasks is 20, EPTD and RPTD-II take 8 ms and 17 ms, respectively, while SATE only takes 0.5 ms. Figure 6.8 shows that as the number of users increases, SATE is more efficient than EPTD but takes more time than RPTD-II. For example, when the number of users is 10, EPTD, RPTD-II, and SATE take 17.64, 0.28, and 2.86 s, respectively. It is because SATE adds additional data encryption and decryption operations to preserve the truth's privacy. Similarly, Fig 6.8 shows that as the number of task types increases, the performance of SATE falls between EPTD and RPTD-II.

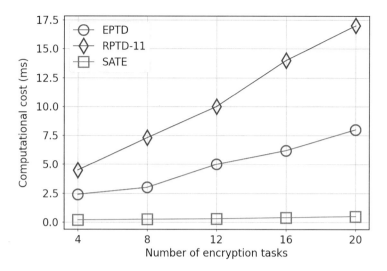

**Fig. 6.6** Distance data collection experiment sensing user computational cost

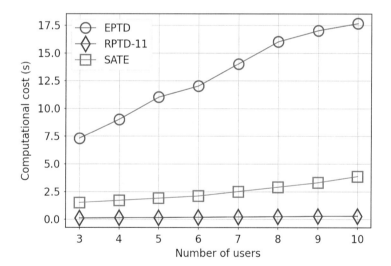

**Fig. 6.7** Change the number of users in the distance data collection experiment mobile crowd-sensing platform computational cost

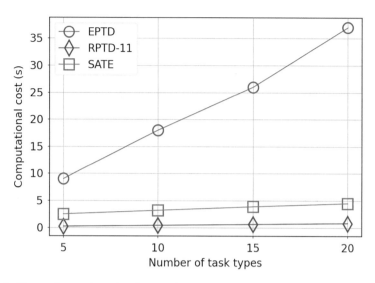

**Fig. 6.8** Change the number of task types in the distance data collection experiment mobile crowdsensing platform computational cost

Then, we design experiments on categorical data. Specifically, we are given several questions, and the user uses a mobile device to randomly select an answer and upload it to the mobile crowdsensing platform. In this experiment, we use 10 mobile phones as sensing users and select 20 questions, each containing 4 optional answers. Each experiment was run 10 times, and the average value was selected as the experimental result. Figures 6.9, 6.10, and 6.11 show the computational cost required for sensing client-side and cloud server-side computations. Since categorical data can be regarded as special continuous data, the computational overheads of Figs. 6.9, 6.11, and 6.11 on the sensing user side and mobile crowdsensing platform side are similar to the results in Figs. 6.9, 6.10, and 6.11.

### 6.5.1.2 Experiments Based on Simulated Mobile Crowdsensing Scenarios

To further verify the computational efficiency of SATE, we randomly generate large-scale sensory data for experiments. Specifically, the data in the synthetic dataset was randomly selected from 20 to 30, the number of tasks from 60 to 100, and the number of users from 100 to 500. Figures 6.12, 6.13, and 6.14 show the computational cost required on the sensing client and cloud server. As can be seen from Fig. 6.12, when the number of tasks is 100, EPTD takes 41 ms, RPTD-II takes 79 ms, and SATE only takes 5.7 ms. Figures 6.13 and 6.14 show the total running time of the mobile crowdsensing platform side's initialization and iteration when changing the number of users and task types, respectively, where the number

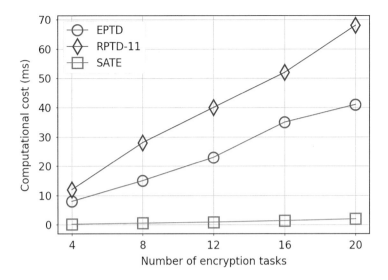

**Fig. 6.9** Question answer collection experiment sensing user computational cost

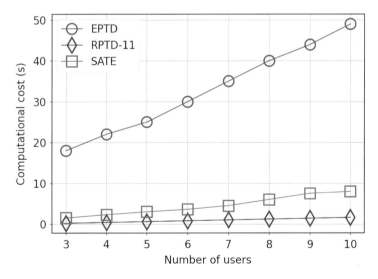

**Fig. 6.10** Change the number of users in the question answer collection experiment mobile crowdsensing platform computational cost

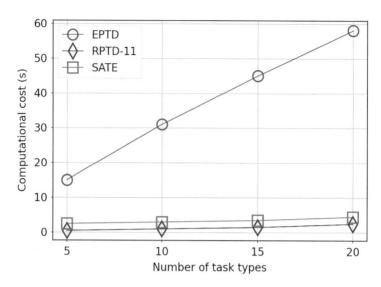

**Fig. 6.11** Change the number of task types in the question answer collection experiment mobile crowdsensing platform computational cost

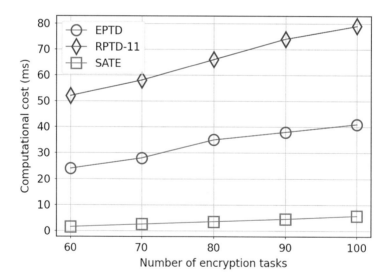

**Fig. 6.12** Synthetic data experiment sensing user computational cost

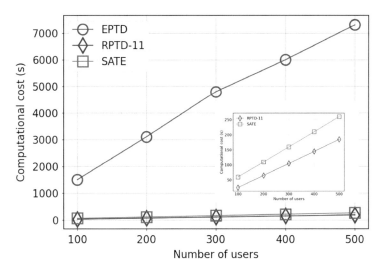

**Fig. 6.13** Change the number of users in the synthetic data experiment mobile crowdsensing platform computational cost

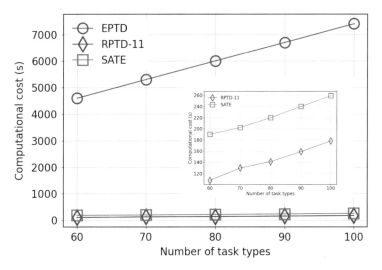

**Fig. 6.14** Change the number of task types in the synthetic data experiment mobile crowdsensing platform computational cost

**Table 6.2** Comparison of communication overhead between SATE and existing solutions

|  |  | SATE | EPTD | RPTD-II |
|---|---|---|---|---|
| Initialization | Data upload | – | $M\|Z\|$ | $(2M+1)\|Z\|$ |
|  | Ciphertext initialization | $(KM+K)\|Z\|$ | – | $KM\|Z\|$ |
| Iteration | Weight update | $(2K+2)\|Z\|$ | $(4KM+2)\|Z_{n^2}\|$ | $(K+1)\|Z\|$ |
|  | Truth update | $(K+3M)\|Z\|$ | $(2M+2)\|Z\|$ | $(K+2M)\|Z\|$ |

of iterations is set to 10. It can be seen that SATE achieves higher computational efficiency among the three schemes.

## 6.5.2   Communication Overhead

We analyze the communication overhead of SATE during system initialization and iteration. Since the size of the random number is negligible compared to the ciphertext, we mainly consider the ciphertext transmission during initialization and iteration. Specifically, we use $|Z|$ to denote the ciphertext size. Table 6.2 shows the comparison of communication overhead between SATE and schemes EPTD and RPTD-II, where $K$ represents the number of sensing users and $M$ represents the number of task types. Due to the use of data perturbation technology in SATE to preserve data privacy, compared with EPTD and RPTD-II, the amount of data transmission can be greatly reduced. The communication overhead of SATE on the mobile crowdsensing platform is less than that of EPTD but more than that of RPTD-II. This is because RPTD-II ignores the privacy-preserving of the truth, thus reducing the transmission of ciphertext related to the truth.

## 6.6   Summary

In this chapter, we propose a privacy-preserving truth discovery scheme SATE for the mobile crowdsensing scenarios with truth hiding. SATE can effectively preserve the truth privacy of data requesters and the sensory data and weight privacy of sensing users. In addition, due to the application of data perturbation technology on the sensing user side, SATE can significantly reduce the computational cost and communication overhead of sensing users. Based on this chapter, it is a meaningful research direction to study the blockchain-based privacy-preserving truth discovery scheme to realize the incentive mechanism based on truth discovery.

# References

1. Li, Y., Li, Q., Gao, J., Su, L., Zhao, B., Fan, W., Han, J.: On the discovery of evolving truth. In: Proceedings of the 21th ACM International Conference on Information and Knowledge Management, pp. 675–684 (2015)
2. Ma, F., Li, Y., Li, Q., Qiu, M., Gao, J., Zhi, S., Su, L., Zhao, B., Ji, H., Han, J.: Faitcrowd: fine grained truth discovery for crowdsourced data aggregation. In: Proceedings of the 21th ACM SIGKDD International Conference on Knowledge Discovery and Data Mining, pp. 745–754 (2015)
3. Yin, X., Han, J., Yu, P.S.: Truth discovery with multiple conflicting information providers on the web. IEEE Trans. Knowl. Data Eng. **20**(6), 796–808 (2008)
4. Miao, C., Su, L., Jiang, W., Li, Y., Tian, M.: A lightweight privacy-preserving truth discovery framework for mobile crowd sensing systems. In: 2017 IEEE Conference on Computer Communications, INFOCOM 2017, Atlanta, GA, USA, May 1–4, 2017, pp. 1–9. IEEE (2017)
5. Tang, J., Fu, S., Xu, M., Luo, Y., Huang, K.: Achieve privacy-preserving truth discovery in crowdsensing systems. In: Proceedings of the 28th ACM International Conference on Information and Knowledge Management, pp. 1301–1310 (2019)
6. Tang, X., Wang, C., Yuan, X., Wang, Q.: Non-interactive privacy-preserving truth discovery in crowd sensing applications. In: IEEE Conference on Computer Communications (INFOCOM), pp. 1988–1996 (2018)
7. Zhang, C., Zhu, L., Xu, C., Liu, X., Sharif, K.: Reliable and privacy-preserving truth discovery for mobile crowdsensing systems. IEEE Trans. Depend. Secure Comput. **18**(3), 1245–1260 (2019)
8. Li, Q., Li, Y., Gao, J., Zhao, B., Fan, W., Han, J.: Resolving conflicts in heterogeneous data by truth discovery and source reliability estimation. In: Proceedings of the 2014 ACM SIGMOD International Conference on Management of Data, pp. 1187–1198 (2014)
9. Li, Y., Li, Q., Gao, J., Su, L., Zhao, B., Fan, W., Han, J.: Conflicts to harmony: a framework for resolving conflicts in heterogeneous data by truth discovery. IEEE Trans. Knowl. Data Eng. **28**(8), 1986–1999 (2016)
10. Liu, X., Qin, B., Deng, R.H., Lu, R., Ma, J.: A privacy-preserving outsourced functional computation framework across large-scale multiple encrypted domains. IEEE Trans. Comput. **65**(12), 3567–3579 (2016)
11. Bresson, E., Catalano, D., Pointcheval, D.: A simple public-key cryptosystem with a double trapdoor decryption mechanism and its applications. In: International Conference on the Theory and Application of Cryptology and Information Security, pp. 37–54 (2003)
12. Dong, C., Wang, Y., Aldweesh, A., McCorry, P., van Moorsel, A.: Betrayal, distrust, and rationality: smart counter-collusion contracts for verifiable cloud computing. In: Proceedings of the 2017 ACM SIGSAC Conference on Computer and Communications Security, pp. 211–227 (2017)
13. Lin, J., Yang, D., Wu, K., Tang, J., Xue, G.: A sybil-resistant truth discovery framework for mobile crowdsensing. In: IEEE 39th International Conference on Distributed Computing Systems (ICDCS), pp. 871–880 (2019)
14. Fan, J., Li, Q., Cao, G.: Privacy-aware and trustworthy data aggregation in mobile sensing. In: IEEE Conference on Communications and Network Security (CNS), pp. 31–39 (2015)

# Chapter 7
# Privacy-Preserving Truth Discovery with Task Hiding

**Abstract** Mobile crowdsensing has emerged as a popular platform for solving many challenging problems by utilizing users' wisdom and resources. Due to the diversity of users, the data provided by different users may vary significantly, and thus it is important to analyze user data quality during data aggregation. Truth discovery has been proven to be an effective mechanism to capture data quality and calculate accurate true values via a weighted combination. In spite of the appealing benefits, existing works on truth discovery either fall short of achieving thorough privacy preservation for participating users or cause tremendous computational and communication overhead. In this chapter, we study challenging problems of truth discovery in mobile crowdsensing and present a lightweight privacy-preserving truth discovery scheme, named LPTD, based on the technologies of secure k-nearest neighbor, data perturbation, and matrix decomposition. Through a detailed analysis, we demonstrate that data privacy and task privacy are well preserved during the whole process. Extensive experiments show that our proposed LPTD has practical performance in terms of accuracy, convergence, computational cost, and communication overhead.

## 7.1 Introduction

This section describes the overview, related works, and preliminary of this chapter.

### 7.1.1 Overview

Mobile crowdsensing has received considerable attention in recent years as it can be utilized to solve many challenging tasks [1, 2]. For example, patients who take new drugs can provide feedback on treatment outcomes to help pharmaceutical companies evaluate the effect of new drugs; drivers who drive on the roads can answer the traffic tasks to help potential drivers make optimal driving plan; and customers who fill out the questionnaires can help companies to provide more accurate recommendation services. In these mobile crowdsensing applications, users contribute their wisdom and resources to answer the tasks published on the mobile crowdsensing platform, which can greatly reduce the financial costs and bring tremendous benefits to various application domains.

In spite of the appealing prospects, the reliability of data provided by different users may vary significantly. In practical mobile crowdsensing applications, some users have sufficient knowledge. Therefore, their data is more convincing, while others who are unfamiliar with the mobile crowdsensing tasks may submit biased and meaningless data. Even worse, some users may even provide deceptive responses driven by malicious or financial purposes. Thus, an important research issue in mobile crowdsensing is: *how to find the accurate true values from the conflicting candidate responses?*

Average or majority voting is an intuitive way to aggregate the data and estimate the accurate results, which has been used in [3–5]. However, these two traditional approaches treat all users equally, which are unable to distinguish the high-quality data from the low-quality ones. Due to the diversity of users, it is essential to capture the difference in user reliability and incorporate the reliability information into truth estimation. However, a challenge that emerges here is that user reliability is usually unknown a priori and it is difficult to estimate user reliability without any supervision. To address this challenge, truth discovery [6–8], as an effective technology to estimate user reliability without prior knowledge, has received considerable attention. The general principle followed in truth discovery approaches is that: if a user holds high reliability, its data is more likely to be selected as the accurate true values; if a user's data is closer to the accurate true values, this user will be assigned a higher weight. Due to its high efficiency, truth discovery has been widely applied in various mobile crowdsensing applications, and many existing works have shown its advantages in user reliability and accurate true value estimation.

**Privacy Concerns** An important issue missing in prior truth discovery schemes is the preservation of user privacy (i.e., data privacy and task privacy): an individual's private personal information may be disclosed from the data it provides or the tasks that it takes an interest in [9]. For example, patients' feedback on new drugs is valuable to evaluate drugs' effects, but patients' health conditions may be disclosed. Reporting traffic conditions is helpful in alleviating traffic congestion, but drivers' locations will be known by other strangers. Customers who fill in questionnaires can help companies provide better recommendation services. However, customers' hobbies may be learned by the companies. Intuitively, we can use encryption primitives to preserve the privacy of users' data, such that no one can know the data in plaintexts. However, since the mobile crowdsensing tasks are usually correlated with users' occupations, education levels, locations, etc., users' sensitive information can still be revealed from the tasks they choose to answer, even if the data are encrypted. For instance, a student may be identified by the courses he/she selects to evaluate. A driver's trajectory may be recovered by the traffic tasks he/she chooses to answer. Thus, both data privacy and task privacy should be preserved in mobile crowdsensing applications. Without good privacy preservation, users may be reluctant to provide their data, and as a result, the flourish of mobile crowdsensing question-answering applications may be impeded.

To achieve privacy-preserving truth discovery, there are some schemes proposed by utilizing various traditional encryption primitives (e.g., garbled circuits, oblivious transfer, homomorphic encryption) [10–13]. These schemes, unfortunately, introduce tremendous computational and communication overhead among the participating entities. Due to the large scale of users involved in mobile crowdsensing applications, the schemes relying on traditional cryptographic tools may not be practical. In addition, to our knowledge, all of the existing schemes fail to consider the preservation of task privacy. In a nutshell, a practical truth discovery scheme that can achieve high efficiency and preserve data privacy and task privacy simultaneously still deserved to be investigated.

**Privacy-Preserving Mechanism** To deal with the above challenges, we propose a lightweight privacy-preserving truth discovery (LPTD) scheme in mobile crowdsensing platforms. This scheme is designed based on the technologies of secure kNN, data perturbation, and matrix decomposition and guarantees both high efficiency and strong privacy preservation (i.e., data privacy and task privacy).

**Contributions** To summarize, the contributions of our scheme are listed below:

- We study challenging problems of truth discovery in mobile crowdsensing applications and identify its utility and security requirements by analyzing users' privacy needs.
- We propose a lightweight privacy-preserving truth discovery scheme, called LPTD, for mobile crowdsensing systems. By exploiting the technology of data perturbation and the properties of matrix multiplications, our proposed scheme can accurately calculate the final true values from the conflicting responses while preserving both data privacy and task privacy.
- Through security analysis, our proposed LPTD can preserve data privacy and task privacy simultaneously. Then, extensive experiments are conducted to demonstrate the accuracy and efficiency of our proposed scheme.

The rest of this chapter is organized as follows: Sect. 7.1.2 reviews the related works on truth discovery; Sect. 7.1.3 provides brief descriptions of majority voting, truth discovery, and secure kNN computation; Sect. 7.2 describes the architecture overview of LPTD from system model, security model, and design goals; Sect. 7.3 presents the detailed design of LPTD; Sect. 7.4 gives the security analysis of LPTD; Sect. 7.5 presents the performance evaluation and analysis; and Sect. 7.6 summarizes the work of this chapter.

## 7.1.2 Related Works

Truth discovery is an effective technology to find truthful information from unreliable sensory data [14–16]. It has been studied much, and many schemes are proposed. The methods used in these schemes can be classified into three categories which are weighted voting-based, bayesian and graphical probabilistic model-based, and optimization-based [17]. Weighted voting-based methods use some variant of majority voting (MV) to find the truthful result [18]. However, MV assumes that each answer has the same quality, which leads to a result that has lower accuracy. Bayesian and graphical probabilistic model-based methods were proposed to deal with the problem caused by MV [19, 20]. Optimization-based methods rely on setting an optimization function [21]. With the iteration process, the optimization function can form a relationship between sources' qualities and claims' true value by computing these two sets of parameters jointly.

To the best of our knowledge, there was no consideration about privacy-preserving in the truth discovery until Miao et al. [22] proposed a secure truth discovery scheme applied to the mobile crowdsensing scenario. So far, a list of privacy-preserving truth discovery schemes was proposed [23–25] (just a few) and various encryption methods have been involved. Miao et al. [22] used homomorphic cryptographic technology to encrypt users' data, which is time-consuming. Differential privacy is leveraged in [26, 27] to preserve users' bid privacy and location privacy in mobile crowdsensing, respectively. Xu et al. [28] designed a perturbation-based scheme, which achieves a better performance than Miao et al. [22]. Zheng et al. proposed two schemes for privacy-preserving truth discovery. At first, Zheng et al. [29] presented a scheme in which users participate in iteration much. So they proposed a new scheme in [30] using garbled circuit and additive homomorphic cryptosystem to reduce user participation. Similar to Zheng et al. [29], Miao et al. [10] presented an efficient and privacy-preserving truth discovery scheme in which the two-cloud model is leveraged. Specifically, Sun et al. [25] presented a novel truth discovery algorithm that not only preserves workers' privacy but also can pay workers based on the workers' privacy cost.

In our work, we do not use traditional encryption methods such as homomorphic cryptosystem appearing in [11, 12, 31]; instead we choose data perturbation [32], secure kNN, and matrix decomposition. We propose a lightweight privacy-preserving truth discovery scheme called LPTD. In LPTD, we preserve data privacy and task privacy simultaneously, which most of the literature [13, 28–30] does not consider.

## 7.1.3 Preliminary

In this section, we give brief descriptions of majority voting, truth discovery, and secure kNN computation before presenting the details of our proposed LPTD.

### 7.1.3.1 Majority Voting

Majority voting is a straightforward way to estimate the truths from the collected answers. It selects the most frequent answer as the aggregated result, which can be mathematically calculated as

$$x_m^* = \arg\max_{x \in \mathcal{X}} \sum_{u_k \in \mathcal{U}} \mathbb{I}(x, x_{k,m}),$$  (7.1)

where $\mathcal{X}$, $\mathbb{I}$ are represented as the set of answers and the indicator function, respectively.

The main drawback of majority voting is that it treats all workers equally, which, however, may not always be true in practical applications. Due to differences in education level, hobbies, religious beliefs, etc., the quality of each worker's answer may vary significantly, resulting in difficulty for this aggregation method to derive accurate results. Thus, a better way to improve the accuracy of the final results is to distinguish the users that have high-quality answers and count these users' answers more when estimating the final results.

### 7.1.3.2 Truth Discovery

Truth discovery is a promising technology to resolve conflicts in noisy data and calculate workers' weight by their submitted data. Many prior works [33–35] have shown its advantages in various scenarios. Specifically, truth discovery starts with the initialization of each task's truth, and the weights and truths are then iteratively updated by the following steps until convergence is satisfied:

- *Weight update.* As the initial ground truth $t_i$ for each task is assumed to be known, each worker's weight $w_k$ will be calculated based on the difference between the worker's data and ground truth. We calculate weights as

$$w_k \leftarrow f\left(\sum_{i=1}^{M} d(x_{k,i}, t_i)\right).$$  (7.2)

In Eq. 7.2, $f(\cdot)$ is a function holding decreasing property, and $d(\cdot)$ is a distance function to measure the difference between the worker's data and ground truth.
- *Truth update.* With each worker's weight calculated by the previous step, the truth for task $i$ can be updated as

$$t_i \leftarrow \frac{\sum_{k=1}^{K} w_k \times x_{k,i}}{\sum_{k=1}^{K} w_k}.$$  (7.3)

### 7.1.3.3   Secure kNN Computation

A secure k-nearest neighbor (kNN) scheme can calculate the inner product of two vectors under encrypted data privately [36, 37]. Specifically, a secure kNN computation usually consists of four parts, i.e., initialization, index generation, trapdoor generation, and query.

- *Initialization.* The key generation center initializes the secret key $\psi = (\mathbf{S}, M_1, M_2)$ randomly, where $\mathbf{S}$ represents an n-dimensional binary vector, which is exploited to split plaintext vectors, and $M_1, M_2$ denote two n×n-dimensional invertible matrixes, which are utilized to encrypt vectors. At last, the key generator center distributes secret key $\psi$ to the requester and the worker through a secure channel.
- *Index generation.* Firstly, the worker generates $n$-dimensional vector $\mathbf{M}_i$ as index, where $i \in [1, K]$. Next, the worker utilizes $\mathbf{S}$ to split $\mathbf{M}_i$ into $\mathbf{M}_{i,a}$ and $\mathbf{M}_{i,b}$ randomly. Specifically, for $j$ from 0 to n-1, if $S(j) = 0$, the worker sets $\mathbf{M}_i[j] = \mathbf{M}_{i,a}[j] = \mathbf{M}_{i,b}[j]$, and if $S(i) = 1$, $\mathbf{M}_i[j]$ is randomly divided into $\mathbf{M}_i[j] = \mathbf{M}_{i,a}[j] + \mathbf{M}_{i,b}[j]$. At last, the worker encrypts the index as $R_i = \{\mathbf{M}_{i,a} \times M_1, \mathbf{M}_{i,b} \times M_2\}$ and submits the encrypted index to the cloud.
- *Trapdoor generation.* Firstly, the requester generates an n-dimensional vector $\mathbf{U}$. Next, the requester utilizes $\mathbf{S}$ to split $\mathbf{U}$ into $\mathbf{U}_a$ and $\mathbf{U}_b$ randomly. Specifically, for $j$ from 0 to n-1, if $S(j) = 0$, the requester randomly sets $\mathbf{U}_a[j], \mathbf{U}_b[j]$ as $\mathbf{U}[j] = \mathbf{U}_a[j] + \mathbf{U}_b[j]$, and if $S(i) = 1$, $\mathbf{U}[j] = \mathbf{U}_a[j] = \mathbf{U}_b[j]$. At last, the requester encrypts the trapdoor as $Q = \{M_1^{-1} \times \mathbf{U}_a, M_2^{-1} \times \mathbf{U}_b\}$ and submits the encrypted trapdoor to the cloud.
- *Query.* When the cloud receives the encrypted index and trapdoor, the cloud calculates

$$
\begin{aligned}
P_i &= R_i \times Q \\
&= \{\mathbf{M}_{i,a} \times M_1, \mathbf{M}_{i,b} \times M_2\} \\
&\quad \times \{M_1^{-1} \times \mathbf{U}_a, M_2^{-1} \times \mathbf{U}_b\} \\
&= \mathbf{M}_{i,a} \times \mathbf{U}_a + \mathbf{M}_{i,b} \times \mathbf{U}_b \\
&= \mathbf{M}_i \times \mathbf{U}.
\end{aligned}
\tag{7.4}
$$

If $P_i$ meets the predetermined conditions, the ciphertext $I_i$ associated with $P_i$ will be returned to the requester.

In secure kNN technology, all users own the same secret keys $\{M_1, M_2\}$. In mobile crowdsensing, users are not fully trusted and will obtain the privacy of others through secret keys. Thus, it is non-trivial to deal with privacy preservation via secure kNN in mobile crowdsensing. In LPTD, we combine secure kNN technology and matrix decomposition technology so that different users have different keys.

## 7.2 Architecture Overview

In this section, we introduce the system model, security model, and design goals of our proposed LPTD.

### 7.2.1 System Model

In our scheme, LPTD involves four entities, i.e., key generator center (KGC), workers, requesters, and cloud. Our proposed system architecture of LPTD is shown in Fig. 7.1.

- KGC: The KGC is a trusted third party that generates and distributes system parameters to other entities in the system.
- Requesters: Requesters are usually organizations and individuals who publish tasks they require to the cloud and obtain the corresponding values from the cloud.
- Workers: Workers are usually individuals who submit the encrypted sensory data to the cloud.

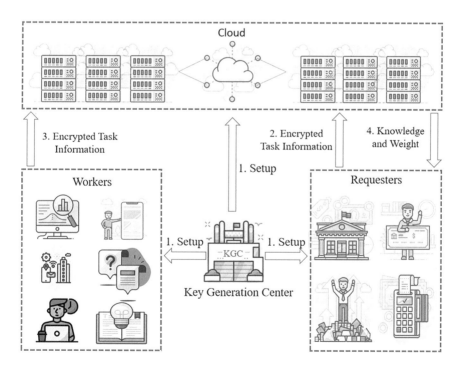

**Fig. 7.1** System architecture of LPTD

- Cloud: The cloud is employed as the mobile crowdsensing platform in the system. It is responsible for receiving the workers' data, performing privacy-preserving truth discovery, and sending the corresponding results to the requester.

At a high level, the process of LPTD is described as follows: the KGC performs system initialization and generates system parameters for requesters, workers, and the cloud; requesters publish the encrypted task information to the cloud; workers submit the encrypted task information to the cloud; the cloud and the requesters iteratively perform the knowledge discovery phase.

## 7.2.2  Security Model

The KGC is a fully trusted third party, and all communications with the KGC are secure. Other entities (i.e., workers, requesters, the cloud) are honest-but-curious. They are honest in performing the designed protocol, but they attempt to obtain sensitive information from other entities. Then, we assume that the cloud would not collude with other entities or pretend to be other valid entities. The assumption for the cloud is reasonable because the cloud is usually a large service worker and understands the importance of reputation [34, 35, 38]. Therefore, we consider the following security models:

- *Ciphertext-only attack model.* The adversary is able to obtain the ciphertexts submitted by workers and requesters, but it has no idea of the plaintexts. This attack occurs when the adversary eavesdrops on the communication channel between the users (workers and requesters) and the cloud. The adversary is aiming to gain some information (e.g., workers' task interests, workers' sensory data, and requesters' task interests) by launching the ciphertext-only attack. The adversary can be each entity in the system.
- *Known-plaintext attack model.* Besides the ciphertexts, the adversary also knows some plaintexts of users' task interest vectors. This attack occurs when the adversary observes an entity's behaviors and eavesdrops on the communication channel among the system entities. The adversary is aiming to gain some information (e.g., workers' sensory data) by launching the known-plaintext attack. The adversary can be each entity in the system.

## 7.2.3  Design Goals

The design goals of our proposed LPTD are as follows:

- **Utility**: LPTD should perform efficient and privacy-preserving truth discovery.
- **Privacy**: Data privacy and task privacy should be preserved under the above security models.

- **Efficiency**: The computational cost and communication overhead introduced on entities should be as low as possible in the system.

## 7.3   Detailed Design

This section introduces the details of LPTD, which mainly contains four entities, i.e., KGC, workers, requesters, and the cloud. As shown in Fig. 7.2, LPTD consists of three phases, i.e., setup, mobile crowdsensing data submission, and privacy-preserving truth discovery. The notations used in this chapter are listed in Table 7.1.

### 7.3.1   Setup

The KGC randomly generates a $(M+4)$-bit vector $\mathbf{S} : S(i)= \{0, 1\}$ and two $(M+4)\times(M+4)$ invertible matrices $\{M_1, M_2\}$. The master secret key is kept as $\mathrm{msk} = \{M_1, M_2, \mathbf{S}\}$. For the users who will participate in our system, they need to register to the KGC. The KGC randomly chooses two $(M+4)\times(M+4)$ invertible matrices $\{A_{i,1}, B_{i,1}\}$ and computes $A_{i,2} = A_{i,1}^{-1} \times M_1$, $B_{i,2} = B_{i,1}^{-1} \times M_2$. Then, it sets the encryption key $ek_i$ and the re-encryption key $rk_i$ as follows:

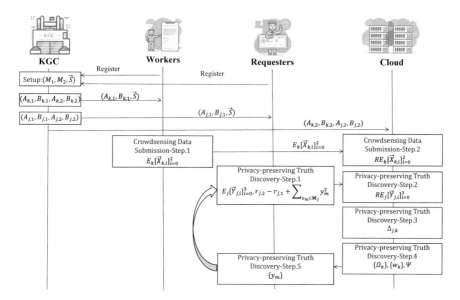

**Fig. 7.2** Workflow of LPTD

**Table 7.1** Notations used in LPTD

| Notation | Meaning |
|---|---|
| M | Maximum number of tasks |
| $\{M_1, M_2, \mathbf{S}\}$ | Master secret key |
| $ek_i = \{A_{i,1}, B_{i,1}, \mathbf{S}\}$ | Encryption secret key of $u_i$ |
| $rk_i = A_{i,2}, B_{i,2}$ | Re-encryption secret key of $u_i$ |
| $\mathbf{l}$ | Task interest vector |
| $o_m$ | $m$-th task |
| $x_{i,m}$ | Sensory data of $u_i$ for $o_m$ |
| $\{\{r_{k,i}\}_{i=1}^{M}, r_{k,a}\}$ | Random values generated by worker $u_k$ |
| $\{r_{j,1}, r_{j,2}, r_{j,3}\}$ | Random values generated by requester $u_j$ |
| $\{\mathbf{X}_{k,0}, \mathbf{X}_{k,1}, \mathbf{X}_{k,2}\}$ | Task information of worker $u_k$ |
| $\{\mathbf{Y}_{j,0}, \mathbf{Y}_{j,1}, \mathbf{Y}_{j,2}, \mathbf{Y}_{j,3}\}$ | Task information of requester $u_j$ |
| $y_m$ | True value of task $o_m$ |
| $w_i$ | Weight of $u_i$ |
| $\mathbb{M}_i$ | Set of tasks that $u_i$ takes interest in |
| $|\mathbb{M}_i|$ | Number of tasks that $u_i$ takes interest in |
| K | Number of workers |
| $E_i[\mathbf{x}]$ | Ciphertext of $\mathbf{x}$ by $ek_i$ |
| $RE_i[\mathbf{x}]$ | Re-encrypted ciphertext of $\mathbf{x}$ by $rk_i$ |
| $\Omega_k$ | Distance between $u_k$'s data and true values |
| $w_k$ | Weight of $u_k$ |
| $\Psi$ | Perturbed weighted true values |

$ek_i = \{A_{i,1}, B_{i,1}, \mathbf{S}\}, rk_i = \{A_{i,2}, B_{i,2}\}$. Finally, the KGC distributes the encryption key $ek_i$ to the user $u_i$, and the re-encryption key $rk_i$ is sent to the cloud via secure channels.

### 7.3.2 Mobile Crowdsensing Data Submission

In this section, each worker sends the encrypted data to the cloud, and the cloud executes re-encryption.

**Step 1** For a worker $u_k$, $u_k$ first generates an M-dimensional task interest vector $\mathbf{l} = (l_{k,1}, l_{k,2}, l_{k,3}, \cdots, l_{k,M})$, where $l_{k,m} = 1$ (or 0) denotes that $u_k$ is interested (or not interested) in the task $o_m$ ($m \in [1, M]$). Then the worker generates an (M+4)-dimensional vector $\mathbf{X}_{k,0} = (0, (1 - l_{k,1}) \cdot r_{k,1}, (1 - l_{k,2}) \cdot r_{k,2}, \cdots, (1 - l_{k,M}) \cdot r_{k,M}, 0, 1, 0)^T$, where $\{r_{k,i}\}_{i=1}^{M}$ are random values. Then, based on the data $\{x_{k,m}\}_{o_m \in \mathbb{M}_k}$ for tasks the worker is interested in, $u_k$ builds two (M+4)-dimensional vectors $\mathbf{X}_{k,1} = (0, a_{k,1}^2, a_{k,2}^2, \cdots, a_{k,M}^2, r_{k,a}, 1, 0)^T$, $\mathbf{X}_{k,2} = (0, a_{k,1}, a_{k,2}, \cdots, a_{k,M}, r_{k,a}, 1, 0)^T$, where $r_{k,a}$ is a random value and $a_{k,m} = x_{k,m}$ (or 0) if $l_{k,m} = 1$ (or 0).

Then, for $\mathbf{x} \in \{\mathbf{X}_{i,0}, \mathbf{X}_{i,1}, \mathbf{X}_{i,2}\}$, the worker splits it into two random vectors as $\{\mathbf{x}_a, \mathbf{x}_b\}$. Specifically, for $i$ from 0 to $M + 3$, if $S(i) = 0$, the worker sets

$$\mathbf{x}_a(i) = \mathbf{x}_b(i) = \mathbf{x}(i), \tag{7.5}$$

and if $S(i) = 1$, the worker randomly sets $\mathbf{x}_a(i)$ and $\mathbf{x}_b(i)$ as

$$\mathbf{x}_a(i) + \mathbf{x}_b(i) = \mathbf{x}(i). \tag{7.6}$$

After that, the worker encrypts $\{\mathbf{x}_a, \mathbf{x}_b\}$ as

$$E_k[\mathbf{x}] = \{A_{k,1}^T \times \mathbf{x}_a, B_{k,1}^T \times \mathbf{x}_b\}, \tag{7.7}$$

where $\mathbf{x} \in \{\mathbf{X}_{k,0}, \mathbf{X}_{k,1}, \mathbf{X}_{k,2}\}$.

Finally, $u_k$ sends the ciphertexts $\{E_k[\mathbf{X}_{k,i}]\}_{i=0}^2$ to the cloud.

**Step 2** In this step, the cloud utilizes its re-encryption keys $(A_{k,2}, B_{k,2})$ to re-encrypt the ciphertexts as

$$\begin{aligned} RE_k[\mathbf{x}] &= \{A_{k,2}^T \times E_k[\mathbf{x}_a], B_{k,2}^T \times E_k[\mathbf{x}_b]\} \\ &= \{M_1^T \times \mathbf{x}_{k,a}, M_2^T \times \mathbf{x}_{k,b}\}, \end{aligned} \tag{7.8}$$

where $\mathbf{x} \in \{\mathbf{X}_{k,0}, \mathbf{X}_{k,1}, \mathbf{X}_{k,2}\}$.

*Remark* In this part, we give the correctness analysis of Eq. 7.8. Recall the equations in the setup, $A_{k,1} \times A_{k,2} = M_1$ and $B_{k,1} \times B_{k,2} = M_2$. According to the basic properties of matrix transpose, we have $(matrix_1 \times matrix_2)^T = matrix_2^T \times matrix_1^T$. Then, we can see that $A_{k,2}^T \times A_{k,1}^T = (A_{k,1} \times A_{k,2})^T = M_1^T$ and $B_{k,2}^T \times B_{k,1}^T = (B_{k,1} \times B_{k,2})^T = M_2^T$.

## 7.3.3  Privacy-Preserving Truth Discovery

In this section, the requester $u_j$ publishes the task to the cloud, and then the requester and the cloud iteratively perform knowledge discovery over encrypted data.

**Step 1** The requester $u_j$ first generates an M-dimensional task interest vector $\mathbf{l} = (l_{j,1}, l_{j,2}, l_{j,3}, \cdots, l_{j,M})$, where $l_{j,m} = 1$ (or 0) denotes that $u_j$ is interested (or not interested) in the task $o_m$ ($m \in [1, M]$). Then, $u_j$ generates two (M+4)-dimensional vectors $\mathbf{Y}_{j,0} = (0, l'_{j,1}, l'_{j,2}, \cdots, l'_{j,M}, 1, 0, 0)^T$, $\mathbf{Y}_{j,1} = (0, l_{j,1}, l_{j,2}, \cdots, l_{j,M}, 1, r_{j,1}, 0)^T$, where $l'_{j,m}$ is a random value (or 0) if $l_{j,m} = 1$ (or 0), and $r_{j,1}$ is a random value.

Next, $u_j$ generates random values $\{y_m\}_{o_m \in \mathbb{M}_j}$ as initialized true values and generates a (M+4)-dimensional vector $\mathbf{Y}_{j,2} = (0, 2b_{j,1}, 2b_{j,2}, \cdots, 2b_{j,M}, 1, r_{j,2}, 0)^T$, where $r_{j,2}$ is a random value and $b_{j,m} = y_m$ (or 0) if $l_{j,m} = 1$ (or 0). In addition, assume $Max_j$ is a large number which is larger than $\max(\sum_{k=1}^{K} w_k x_{k,m})$, where $\sum_{k=1}^{K} w_k x_{k,m}$ is estimated by $u_j$ based on the practical recognition for the tasks. Next, $u_j$ generates random values $\{z_{j,m}\}_{o_m \in \mathbb{M}_j}$, which satisfy $z_{j,1} = 1, \sum_{i=1}^{l-1} z_{j,i} \cdot Max_j < z_{j,l}$ ($l \in [2, |\mathbb{M}_j|]$). Then, $u_j$ builds a (M+4)-dimensional vector $\mathbf{Y}_{j,3} = (0, c_{j,1}, c_{j,2}, \cdots, c_{j,M}, 0, r_{j,3}, 0)^T$, where $c_{j,m} = z_{j,m}$ (or 0) if $l_{j,m} = 1$ (or 0).

Then, for $\mathbf{y} \in \{\mathbf{Y}_{j,0}, \mathbf{Y}_{j,1}, \mathbf{Y}_{j,2}, \mathbf{Y}_{j,3}\}$, the requester splits $\mathbf{y}$ into two random vectors as $\{\mathbf{y}_a, \mathbf{y}_b\}$. Specifically, for $i$ from 0 to $M + 3$, if $S(i) = 1$, the worker sets

$$\mathbf{y}_a(i) = \mathbf{y}_b(i) = \mathbf{y}(i), \tag{7.9}$$

and if $S(i) = 0$, the worker randomly sets $\mathbf{y}_a(i)$ and $\mathbf{y}_b(i)$ as

$$\mathbf{y}_a(i) + \mathbf{y}_b(i) = \mathbf{y}(i). \tag{7.10}$$

Then, the requester executes encryption as

$$E_j[\mathbf{y}] = \{A_{j,1}^{-1} \times \mathbf{y}_a, B_{j,1}^{-1} \times \mathbf{y}_b\}, \tag{7.11}$$

where $\mathbf{y} \in \{\mathbf{Y}_{j,0}, \mathbf{Y}_{j,1}, \mathbf{Y}_{j,2}, \mathbf{Y}_{j,3}\}$. Finally, $u_j$ sends $(\{E_j[\mathbf{Y}_{j,i}]\}_{i=0}^{3}, r_{j,2} - r_{j,1} + \sum_{o_m \in \mathbb{M}_j} y_m^2)$ to the cloud.

**Step 2**  The cloud obtains $(\{E_j[\mathbf{Y}_{j,i}]\}_{i=0}^{3}, r_{j,2} - r_{j,1} + \sum_{o_m \in \mathbb{M}_j} y_m^2)$ published by $u_j$ and utilizes its re-encryption key set $\{A_{j,2}, B_{j,2}\}$ to re-encrypt the ciphertexts. The cloud calculates

$$\begin{aligned} \mathsf{RE}_j[\mathbf{y}] &= \{A_{j,2}^{-1} \times E_j[\mathbf{y}_a], B_{j,2}^{-1} \times E_j[\mathbf{y}_b]\} \\ &= \{M_1^{-1} \times \mathbf{y}_{j,a}, M_2^{-1} \times \mathbf{y}_{j,b}\}, \end{aligned} \tag{7.12}$$

where $\mathbf{y} \in \{\mathbf{Y}_{j,0}, \mathbf{Y}_{j,1}, \mathbf{Y}_{j,2}, \mathbf{Y}_{j,3}\}$.

*Remark*  In this part, we give the correctness analysis of Eq. 7.12. Recall the equations in the setup, $A_{j,1} \times A_{j,2} = M_1$ and $B_{j,1} \times B_{j,2} = M_2$. According to the basic properties of matrix inversion, we have $A_{j,2}^{-1} \times A_{j,1}^{-1} = M_1^{-1}$ and $B_{j,2}^{-1} \times B_{j,1}^{-1} = M_2^{-1}$.

**Step 3** Then, for each worker, the cloud computes

$$
\begin{aligned}
\Delta_{j,k} &= \mathsf{RE}_j[\mathbf{Y}_{j,0}]^T \times \mathsf{RE}_k[\mathbf{X}_{k,0}] \\
&= \{M_1^{-1} \times \mathbf{Y}_{j,0,a}, M_2^{-1} \times \mathbf{Y}_{j,0,b}\}^T \\
&\quad \times \{M_1^T \times \mathbf{X}_{k,0,a}, M_2^T \times \mathbf{X}_{k,0,b}\} \\
&= \mathbf{Y}_{j,0,a}^T \circ \mathbf{X}_{k,0,a} + \mathbf{Y}_{j,0,b}^T \circ \mathbf{X}_{k,0,b} \\
&= \mathbf{Y}_{j,0}^T \circ \mathbf{X}_{k,0}.
\end{aligned}
\tag{7.13}
$$

In Eq. 7.13, o is utilized to denote the inner product production of two vectors. It is obvious that $\Delta_{j,k} = 0$ if $\mathbb{M}_j \in \mathbb{M}_k$. The cloud gets the workers who meet $\Delta_{j,k} = 0$ and we assume that there are $K$ workers who meet $\Delta_{j,k} = 0$. Next, the cloud records these workers' information into a task result list. Then, the cloud performs **Step 4**.

*Remark* We then give the correctness analysis of Eq. 7.13. According to the basic properties of matrix inversion and matrix transpose, it can be seen that $\{M_1^{-1} \times \mathbf{Y}_{j,0,a}\}^T \times \{M_1^T \times \mathbf{X}_{k,0,a}\} = \mathbf{Y}_{j,0,a}^T \times (M_1^{-1})^T \times M_1^T \times \mathbf{X}_{k,0,a} = \mathbf{Y}_{j,0,a}^T \times (M_1^T)^{-1} \times M_1^T \times \mathbf{X}_{k,0,a} = \mathbf{Y}_{j,0,a}^T \circ \mathbf{X}_{k,0,a}$ and $\{M_2^{-1} \times \mathbf{Y}_{j,0,b}\}^T \times \{M_2^T \times \mathbf{X}_{k,0,b}\} = \mathbf{Y}_{j,0,b}^T \times (M_2^{-1})^T \times M_2^T \times \mathbf{X}_{k,0,b} = \mathbf{Y}_{j,0,b}^T \times (M_2^T)^{-1} \times M_2^T \times \mathbf{X}_{k,0,b} = \mathbf{Y}_{j,0,b}^T \circ \mathbf{X}_{k,0,b}$. Then, we assume that workers and requesters have the same task interest vector and set M-dimensional task interest vector $\mathbf{l} = (0, 1, 1, 1, 0, \cdots, 0)$. Then, we have $\mathbf{X}_{k,0} = (0, r_{k,1}, 0, 0, 0, r_{k,5}, r_{k,6}, \cdots, r_{k,M}, 0, 1, 0)^T$ and $\mathbf{Y}_{j,0} = (0, 0, l'_{j,2}, l'_{j,3}, l'_{j,4}, 0, 0, \cdots, 0, 1, 0, 0)^T$, where $\{r_{k,i}\}_{i=1}^M$, $\{l'_{j,i}\}_{i=1}^M$ are random values. Then, we can see $\mathbf{Y}_{j,0}^T \circ \mathbf{X}_{k,0} = \sum_{i=0}^{M+3} Y_{j,0,i} \times X_{k,0,i} = 0$.

**Step 4** The cloud further calculates

$$
\begin{aligned}
\Omega_k &= \mathsf{RE}_j[\mathbf{Y}_{j,1}]^T \times \mathsf{RE}_k[\mathbf{X}_{k,1}] - \mathsf{RE}_j[\mathbf{Y}_{j,2}]^T \times \mathsf{RE}_k[\mathbf{X}_{k,2}] \\
&\quad + r_{j,2} - r_{j,1} + \sum_{o_m \in \mathbb{M}_j} y_m^2 \\
&= \{M_1^{-1} \times \mathbf{Y}_{j,1,a}, M_2^{-1} \times \mathbf{Y}_{j,1,b}\}^T \times \{M_1^T \times \mathbf{X}_{k,1,a}, \\
&\quad M_2^T \times \mathbf{X}_{k,1,b}\} - \{M_1^{-1} \times \mathbf{Y}_{j,2,a}, M_2^{-1} \times \mathbf{Y}_{j,2,b}\}^T \times \\
&\quad \{M_1^T \times \mathbf{X}_{k,2,a}, M_2^T \times \mathbf{X}_{k,2,b}\} + r_{j,2} - r_{j,1} + \sum_{o_m \in \mathbb{M}_j} y_m^2 \\
&= \mathbf{Y}_{j,1}^T \circ \mathbf{X}_{k,1} - \mathbf{Y}_{j,2}^T \circ \mathbf{X}_{k,2} + r_{j,2} - r_{j,1} + \sum_{o_m \in \mathbb{M}_j} y_m^2 \\
&= \sum_{o_m \in \mathbb{M}_j} (x_{k,m} - y_m)^2
\end{aligned}
\tag{7.14}
$$

to obtain the distance between the worker $u_k$'s data and the true values. Then, $u_k$'s weight can be computed as $w_k = \log(\frac{\sum_{i=1}^{K} \Omega_i}{\Omega_k})$.

Then, the cloud computes

$$
\begin{aligned}
\Psi &= \mathsf{RE}_j[\mathbf{Y}_{j,3}]^T \times \sum_{k=1}^{K} w_k \times \mathsf{RE}_k[\mathbf{X}_{k,2}] \\
&= \{M_1^{-1} \times \mathbf{Y}_{j,3,a}, M_2^{-1} \times \mathbf{Y}_{j,3,b}\}^T \times \\
&\quad \sum_{k=1}^{K} w_k \times \{M_1^T \times \mathbf{X}_{k,2,a}, M_2^T \times \mathbf{X}_{k,2,b}\} \\
&= \mathbf{Y}_{j,3}^T \circ \sum_{k=1}^{K} w_k \times \mathbf{X}_{k,2} \\
&= r_{j,3} \cdot \sum_{k=1}^{K} w_k + \sum_{o_m \in \mathbb{M}_j} \sum_{k=1}^{K} w_k \times x_{k,m} \times z_{j,m}
\end{aligned}
\tag{7.15}
$$

to obtain the perturbed weighted true values. Then, the cloud sends $(\Psi, \sum_{k=1}^{K} w_k)$ to the requester.

**Step 5** The requester $u_j$ obtains $(\Psi, \sum_{k=1}^{K} w_k)$ from the cloud and calculates $\Psi_{|\mathbb{M}_j|} = \Psi - r_{j,3} \cdot \sum_{k=1}^{K} w_k$ to obtain the weighted true values. Then, the requester $u_j$ iteratively performs Eq. 7.16 to recover the true value for task $o_m$, where $m \in [2, |\mathbb{M}_j|]$.

$$
\begin{aligned}
y_m &= \frac{\Psi_m - \Psi_m \bmod z_{j,m}}{z_{j,m} \times \sum_{k=1}^{K} w_k}. \\
&= \frac{\Psi_m - \sum_{o_m \in \mathbb{M}_j} \sum_{k=1}^{K} w_k \times x_{k,m} \times z_{j,m} \bmod z_{j,m}}{z_{j,m} \times \sum_{k=1}^{K} w_k} \\
&= \frac{\sum_{k=1}^{K} w_k \times x_{k,m} \times z_{j,m}}{z_{j,m} \times \sum_{k=1}^{K} w_k}.
\end{aligned}
\tag{7.16}
$$

$$\Psi_{m-1} = \Psi_m \bmod z_{j,m}.$$

Then, we have $y_1 = \Psi_1$ for task $o_1$.

Note that for the first time, **Steps 1** and **3** will be completely conducted by the cloud and the requester. In the following iterations, the requester only updates $(\mathsf{E}_j[\mathbf{Y}_{j,2}], r_{j,2} - r_{j,1} + \sum_{o_m \in \mathbb{M}_j} y_m^2)$ to the cloud and the cloud does not perform **Step 3**. Until the distance between two consecutive weight data $\sum_{k=1}^{K} w_k)$ is less than a threshold value, the cloud proposes a service terminal request to the requester with the last $(\Psi, \sum_{k=1}^{K} w_k)$. After the requester obtains the last $(\Psi, \sum_{k=1}^{K} w_k)$, $u_j$ performs **Step 5** to determine whether the distance between two consecutive estimated true values is less than a threshold value. If the distance meets the requirement, the requester sets $y_m$ as the truthful knowledge for the task $o_m$. Otherwise, the requester can apply for arbitration with the KGC.

In the system, the design goals of LPTD are well satisfied. As shown in Sect. 7.4, LPTD preserves the data privacy and task privacy during the whole procedure simultaneously. In LPTD, the workers are not involved in knowledge discovery, and only a few matrices are required to be transmitted between the requesters and the cloud. Thus, very little computational and communication overhead is incurred on the user side. As shown in Sect. 7.5, LPTD has acceptable efficiency in practice. Thus, the utility of LPTD is satisfied.

## 7.4  Security Analysis

In this section, we analyze the security of our proposed (re-)encryption methods and prove the security of LPTD. Since the (re-)encryption methods of workers and requesters are similar, we take the workers' (re-)encryption process as an example to analyze the security of our proposed (re-)encryption methods. In addition, since the known-plaintext attack is more powerful than the ciphertext-only attack, we will focus on analyzing the security of LPTD under the known-plaintext attack model.

**Theorem 1** *Our proposed (re-)encryption methods are secure under the known-plaintext attack model if the adversary cannot obtain the randomizable vector* $\mathbf{S}$.

***Proof Sketch*** We first analyze the security of our proposed encryption method. Under the known-plaintext attack model, the adversary knows the ciphertexts $\{\mathsf{E}_k[\mathbf{X}_{k,i,a}], \mathsf{E}_k[\mathbf{X}_{k,i,b}]\}_{i=0}^{2}$ and the corresponding task interest vector $\mathbf{l}$. We take $\{\mathsf{E}_k[\mathbf{X}_{k,0,a}], \mathsf{E}_k[\mathbf{X}_{k,0,b}]\}$ as an example to analyze the security of our proposed encryption method. Although the adversary knows the task interest vector $\mathbf{l}$, he/she cannot compute the vectors $\{\mathbf{X}_{k,0,a}, \mathbf{X}_{k,0,b}\}$, since he/she cannot obtain the randomizable vector $\mathbf{S}$ [37]. Thus, the adversary can only set $\{\mathbf{X}_{k,0,a}, \mathbf{X}_{k,0,b}\}$ as two random (M+4)-dimensional vectors.

Next, because the adversary cannot obtain the encryption secret key set $\{A_1, B_1\}$, he/she can only set $\{A_1, B_1\}$ as two random (M+4)×(M+4)-dimensional invertible matrices.

Then, the adversary creates two linear equations as follows: $A_1^T \times \mathbf{X}_{k,0,a} = \mathsf{E}_k[\mathbf{X}_{k,1,a}]$ and $B_1^T \times \mathbf{X}_{k,0,b} = \mathsf{E}_k[\mathbf{X}_{k,0,b}]$. There are $2 \times$(M+4) unknown variables in $\mathbf{X}_{k,0,a}$ and $\mathbf{X}_{k,0,b}$, and $2 \times$(M+4)×(M+4) unknown variables in $A_1$ and $B_1$. Since there are only $2 \times$(M+4) equations, which are less than the number of the unknown variables, the adversary cannot solve the equations.

Thus, our proposed encryption method is secure under the known-plaintext attack model. Since the proof of our proposed re-encryption method is similar to the encryption method, we can prove the security of the re-encryption method in the same way. Therefore, Theorem 1 is proven.

**Theorem 2** *If Theorem 1 is proven, the data privacy is well safeguarded in LPTD.*

**Proof Sketch** Since workers do not receive data from the cloud, we only need to demonstrate that the cloud and requesters cannot obtain sensitive data during the whole procedure.

**Case 1** Firstly, we take the cloud as the adversary who is aiming to break the data privacy of LPTD.

In the setup, the cloud can obtain the re-encryption key set $\{A_{i,2}, B_{i,2}\}$. Since the cloud does not know the master key set $\{M_1, M_2\}$ and the $u_i$'s encryption key set $\{A_{i,1}, B_{i,1}\}$, it cannot obtain sensitive data from the re-encryption key set $\{A_{i,2}, B_{i,2}\}$.

In the mobile crowdsensing data submission phase, the cloud can obtain the ciphertexts $\{E_k[\mathbf{X}_{k,i,a}], E_k[\mathbf{X}_{k,i,b}]\}_{i=0}^2$ and $\{RE_k[\mathbf{X}_{k,i,a}], RE_k[\mathbf{X}_{k,i,b}]\}_{i=0}^2$. Since Theorem 1 is proven, the cloud cannot obtain sensitive data from the ciphertexts $\{E_k[\mathbf{X}_{k,i,a}], E_k[\mathbf{X}_{k,i,b}]\}_{i=0}^2$ and $\{RE_k[\mathbf{X}_{k,i,a}], RE_k[\mathbf{X}_{k,i,b}]\}_{i=0}^2$.

In the privacy-preserving truth discovery phase, the cloud can know the ciphertexts $\{E_j[\mathbf{Y}_{j,i,a}], E_j[\mathbf{Y}_{j,i,b}]\}_{i=0}^3$ and $\{RE_j[\mathbf{Y}_{j,i,a}], RE_j[\mathbf{Y}_{j,i,b}]\}_{i=0}^3$. Meanwhile, the cloud can obtain the plaintexts $r_{j,2} - r_{j,1} + \sum_{o_m \in \mathbb{M}_j} y_m^2$, $\Delta_{j,k}$, $w_k$, and $\Psi$. Since Theorem 1 is proven, the cloud cannot obtain sensitive data from the ciphertexts $\{E_j[\mathbf{Y}_{j,i,a}], E_j[\mathbf{Y}_{j,i,b}]\}_{i=0}^3$ and $\{RE_j[\mathbf{Y}_{j,i,a}], RE_j[\mathbf{Y}_{j,i,b}]\}_{i=0}^3$. Since $\{r_{k,i}\}_{i=1}^M$, $r_{j,1}$, $r_{j,2}$, $r_{j,3}$, and $y_m$ are unknown, the cloud cannot obtain sensitive data for workers and requesters from known plaintexts $r_{j,2} - r_{j,1} + \sum_{o_m \in \mathbb{M}_j} y_m^2$, $\Delta_{j,k}$, and $\Psi$. Since the cloud cannot know the users' task interest vectors $\mathbf{l}$, it cannot obtain sensitive information from $w_k$.

**Case 2** We take a specific requester as the adversary, who is aiming to break the data privacy of LPTD.

In the setup phase, the requester can obtain the encryption key set $\{A_{i,1}, B_{i,1}\}$. Since the requester does not know the master key set $\{M_1, M_2\}$ and the re-encryption key set $\{A_{i,2}, B_{i,2}\}$, the requester cannot obtain sensitive data from the encryption key set $\{A_{i,1}, B_{i,1}\}$.

In the privacy-preserving truth discovery phase, the requester only knows the plaintexts $\Psi$ and $\sum_{k=1}^K w_k$. Since the requester cannot obtain the number of workers, who participate in this task, the requester cannot obtain the worker's sensitive data from the summations. Since the requester does not participate in other phases, the requester cannot also obtain sensitive data. Thus, Theorem 2 is proven.

**Theorem 3** *If Theorem 1 is proven, the task privacy is well safeguarded in LPTD.*

**Proof Sketch** We take the cloud as the adversary who is aiming to break the task privacy of LPTD. In LPTD, the (re-)encryption methods of task information are similar to those used for sensory data. Since Theorem 2 is proved, the cloud cannot obtain sensitive data from all phases in LPTD. Similarly, the cloud cannot obtain sensitive task information during the whole procedure.

## 7.5  Performance Evaluation and Analysis

In this section, we evaluate the performance of our proposed LPTD based on practical mobile crowdsensing systems and compare our scheme with CRH [15], RPTD [39], and $L^2$-PPTD [40]. In comparison, we implement the baseline truth discovery mechanism CRH for accuracy comparison. Then, we implement the two most recent efficient and privacy-preserving truth discovery schemes RPTD and $L^2$-PPTD for efficiency comparison. Finally, we evaluate the blockchain performance of our proposed LPTD.

**Experimental Configuration**  In our experiments, we take the transport mode recognition system as an example of numerical data to illustrate the performance of LPTD. Specifically, our scheme is implemented in Java and Ethereum programming language Solidity. A laptop with 16 GB of RAM and 1.80 GB GHz 8th generation Intel Core is employed as the cloud and requesters, and smartphones with 6.0 GB RAM are employed as workers. The smart contract is deployed on a Javascript VM environment. In addition, the sensory data is composed of floating-point numbers, and we regard the daily traffic speed statistics event as a task. For cryptographic parameters, we set the public-key size as 512 bits.

### 7.5.1  Accuracy

We use the standard root of mean squared error (RMSE) between the estimated truth value and the ground true value to measure the accuracy of our scheme. We measure the accuracy by varying the number of workers and comparing it with that of CRH. The comparison result is shown in Fig. 7.3. In Fig. 7.3, the number of tasks is set as 20, and the number of workers ranges from 2 to 10. It is observed that our scheme can achieve the same estimation accuracy as CRH regardless of the worker number.

### 7.5.2  Convergence

We evaluate the ability of convergence in different iterations between LPTD and CRH and plot the results in Fig. 7.4. Specifically, we randomly select a task and set the number of data nodes as 10. It is observed that our scheme converges quickly in a few iterations and has a similar convergence pattern to CRH.

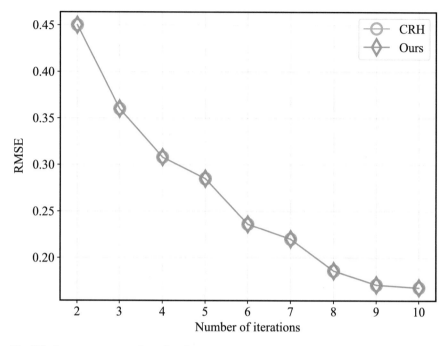

**Fig. 7.3**  Accuracy w.r.t. number of workers

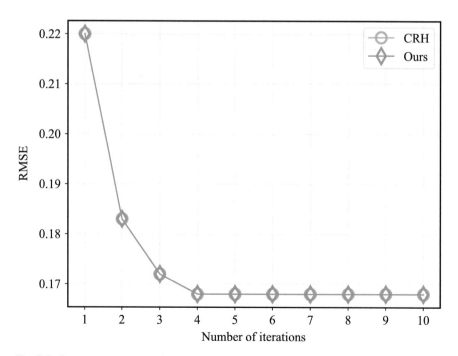

**Fig. 7.4**  Convergence w.r.t. number of iterations

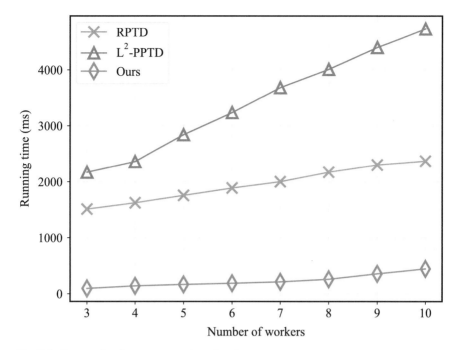

**Fig. 7.5** Computational cost w.r.t. number of workers

### 7.5.3 Computational Cost

We measure the computational cost of our scheme. We evaluate the computational cost for truth discovery on the cloud side and compare it with that of RPTD and $L^2$-PPTD. The comparison results are shown in Figs. 7.5 and 7.6. In Fig. 7.5, the number of tasks is set as 20, and the number of workers ranges from 3 to 10. When the number of workers equals 10, our scheme needs 445 ms to finish the whole privacy-preserving truth discovery procedure, which is more efficient than RPTD and $L^2$-PPTD.

In Fig. 7.6, the number of workers is set as 10, and the number of tasks varies from 4 to 20. Our scheme needs less running time to finish the whole privacy-preserving truth discovery procedure than the other two schemes. The reason is that RPTD and $L^2$-PPTD rely on the traditional cryptographic tool to preserve privacy, causing tremendous computational costs. Our scheme relies on secure kNN

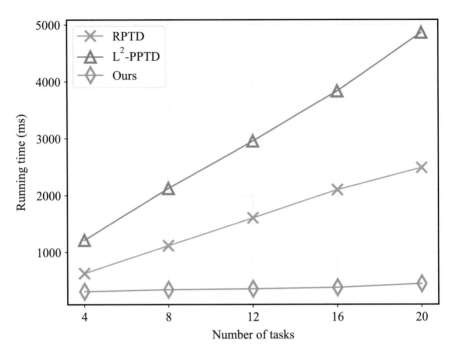

**Fig. 7.6** Computational cost w.r.t. number of tasks

computation, matrix decomposition, and data perturbation, which are regarded as
lightweight algorithms.

### 7.5.4   Communication Overhead

At last, we evaluate the communication overhead on the worker side and compare
it with that of RPTD and $L^2$-PPTD. The comparison results are shown in Fig. 7.7.
From the figure, we can see that with the increase in task number, the communication overhead of LPTD is more than that of $L^2$-PPTD but is less than that of RPTD.
The reason is that $L^2$-PPTD transmits plaintexts of perturbed data and perturbations
to the cloud, while LPTD transmits six matrices to the cloud. However, since the
workers are not involved in knowledge discovery, we emphasize the communication
overhead on the worker side is a one-time cost. The results demonstrate that LPTD
has acceptable efficiency in practice.

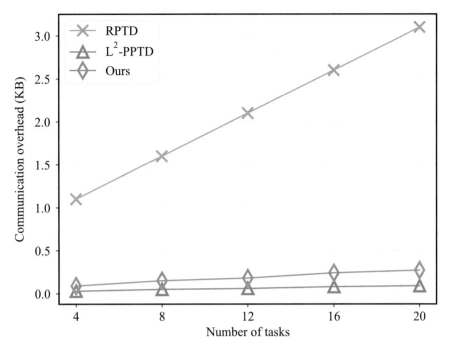

**Fig. 7.7** Communication overhead w.r.t. number of tasks

## 7.6 Summary

In this chapter, we propose a lightweight privacy-preserving truth discovery scheme, called LPTD, based on secure kNN, data perturbation, and matrix decomposition. In LPTD, the workers are not involved in knowledge discovery, and only a few matrices are required to be transmitted between the requesters and the cloud. Thus, very little computational cost and communication overhead are incurred on the user side. Formal analysis shows that LPTD can preserve data privacy and task privacy under the known-plaintext attack simultaneously. We finally conduct extensive experiments to demonstrate that LPTD has a practical performance in terms of accuracy, convergence, computational cost, and communication overhead.

For future work, we will design a truth discovery scheme that is secure under the chosen-plaintext attack and chosen-ciphertext attack. Then, we will focus on the trustworthiness of the cloud and try to design a verifiable efficient and privacy-preserving truth discovery scheme for mobile crowdsensing.

# References

1. Tu, J., Cheng, P., Chen, L.: Quality-assured synchronized task assignment in crowdsourcing. IEEE Trans. Knowl. Data Eng. **33**(3), 1156–1168 (2021)
2. Zhang, X., Wu, Y., Huang, L., Ji, H., Cao, G.: Expertise-aware truth analysis and task allocation in mobile crowdsourcing. IEEE Trans. Mob. Comput. **20**(3), 1001–1016 (2021)
3. Sheng, V.S., Provost, F., Ipeirotis, P.G.: Get another label? Improving data quality and data mining using multiple, noisy labelers. In: Proceedings of the 14th ACM SIGKDD International Conference on Knowledge Discovery and Data Mining, Las Vegas, Nevada, USA, August 24–27, 2008, pp. 614–622 (2008)
4. Smyth, P., Burl, M.C., Fayyad, U.M., Perona, P.: Knowledge discovery in large image databases: dealing with uncertainties in ground truth. In: Knowledge Discovery in Databases: Papers from the 1994 AAAI Workshop, Seattle, Washington, USA, July 1994. Technical Report WS-94-03, pp. 109–120 (1994)
5. Donmez, P., Carbonell, J.G., Schneider, J.G. Efficiently learning the accuracy of labeling sources for selective sampling. In: Proceedings of the 15th ACM SIGKDD International Conference on Knowledge Discovery and Data Mining, Paris, France, June 28–July 1, 2009, pp. 259–268 (2009)
6. Li, Y., Li, Q., Gao, J., Su, L., Zhao, B., Fan, W., Han, J.: On the discovery of evolving truth. In: Proceedings of the 21th ACM SIGKDD International Conference on Knowledge Discovery and Data Mining, Sydney, NSW, Australia, August 10–13, 2015, pp. 675–684 (2015)
7. Ma, F., Li, Y., Li, Q., Qiu, M., Gao, J., Zhi, S., Su, L., Zhao, B., Ji, H., Han, J.: Faitcrowd: fine grained truth discovery for crowdsourced data aggregation. In: Proceedings of the 21th ACM SIGKDD International Conference on Knowledge Discovery and Data Mining, Sydney, NSW, Australia, August 10–13, 2015, pp. 745–754 (2015)
8. Yin, X., Han, J., Yu, P.S.: Truth discovery with multiple conflicting information providers on the web. IEEE Trans. Knowl. Data Eng. **20**(6), 796–808 (2008)
9. Xue, L., Ni, J., Huang, C., Lin, X., Shen, X.: Forward secure and fine-grained data sharing for mobile crowdsensing. In: 17th International Conference on Privacy, Security and Trust, PST 2019, Fredericton, NB, Canada, August 26–28, 2019, pp. 1–9 (2019)
10. Miao, C., Su, L., Jiang, W., Li, Y., Tian, M.: A lightweight privacy-preserving truth discovery framework for mobile crowd sensing systems. In: INFOCOM, pp. 1–9 (2017)
11. Tang, X., Wang, C., Yuan, X., Wang, Q.: Non-interactive privacy-preserving truth discovery in crowd sensing applications. In: INFOCOM, pp. 1988–1996. IEEE (2018)
12. Tang, J., Fu, S., Xu, M., Luo, Y., Huang, K.: Achieve privacy-preserving truth discovery in crowdsensing systems. In: CIKM, pp. 1301–1310 (2019)
13. Zhang, C., Zhu, L., Xu, C., Liu, X., Sharif, K.: Reliable and privacy-preserving truth discovery for mobile crowdsensing systems. IEEE Trans. Depend. Secure Comput. **18**(3), 1245–1260 (2019)
14. Li, Q., Li, Y., Gao, J., Zhao, B., Fan, W., Han, J.: Resolving conflicts in heterogeneous data by truth discovery and source reliability estimation. In: International Conference on Management of Data, SIGMOD 2014, Snowbird, UT, USA, June 22–27, 2014, pp. 1187–1198 (2014)
15. Li, Y., Li, Q., Gao, J., Su, L., Zhao, B., Fan, W., Han, J.: Conflicts to harmony: a framework for resolving conflicts in heterogeneous data by truth discovery. IEEE Trans. Knowl. Data Eng. **28**(8), 1986–1999 (2016)
16. Zheng, Y., Li, G., Li, Y., Shan, C., Cheng, R.: Truth inference in crowdsourcing: is the problem solved? Proc. VLDB Endow. **10**(5), 541–552 (2017)
17. Berti-Équille, L.: Truth discovery. In: Encyclopedia of Big Data Technologies (2019)
18. Galland, A., Abiteboul, S., Marian, A., Senellart, P.: Corroborating information from disagreeing views. In: Proceedings of the Third International Conference on Web Search and Web Data Mining, WSDM 2010, New York, NY, USA, February 4–6, 2010, pp. 131–140 (2010)

19. Zhi, S., Zhao, B., Tong, W., Gao, J., Yu, D., Ji, H., Han, J.: Modeling truth existence in truth discovery. In: Proceedings of the 21th ACM SIGKDD International Conference on Knowledge Discovery and Data Mining, Sydney, NSW, Australia, August 10–13, 2015, pp. 1543–1552 (2015)

20. Waguih, D.A., Goel, N., Hammady, H.M., Berti-Équille, L.: Allegatortrack: combining and reporting results of truth discovery from multi-source data. In: 31st IEEE International Conference on Data Engineering, ICDE 2015, Seoul, South Korea, April 13–17, 2015, pp. 1440–1443 (2015)

21. Marshall, J., Argueta, A., Wang, D.: A neural network approach for truth discovery in social sensing. In: 14th IEEE International Conference on Mobile Ad Hoc and Sensor Systems, MASS 2017, Orlando, FL, USA, October 22–25, 2017, pp. 343–347 (2017)

22. Miao, C., Jiang, W., Su, L., Li, Y., Guo, S., Qin, Z., Xiao, H., Gao, J., Ren, K.: Cloud-enabled privacy-preserving truth discovery in crowd sensing systems. In: Proceedings of the 13th ACM Conference on Embedded Networked Sensor Systems, pp. 183–196 (2015)

23. Tang, X., Wang, C., Yuan, X., Wang, Q.: Non-interactive privacy-preserving truth discovery in crowd sensing applications. In: 2018 IEEE Conference on Computer Communications, INFOCOM 2018, Honolulu, HI, USA, April 16–19, 2018, pp. 1988–1996 (2018)

24. Zheng, Y., Duan, H., Wang, C.: Learning the truth privately and confidently: encrypted confidence-aware truth discovery in mobile crowdsensing. IEEE Trans. Inf. Forensics Secur. **13**(10), 2475–2489 (2018)

25. Sun, P., Wang, Z., Feng, Y., Wu, L., Li, Y., Qi, H., Wang, Z.: Towards personalized privacy-preserving incentive for truth discovery in crowdsourced binary-choice question answering. In: 39th IEEE Conference on Computer Communications, INFOCOM 2020, Toronto, ON, Canada, July 6–9, 2020, pp. 1133–1142 (2020)

26. Jin, H., Su, L., Ding, B., Nahrstedt, K., Borisov, N.: Enabling privacy-preserving incentives for mobile crowd sensing systems. In: 36th IEEE International Conference on Distributed Computing Systems, ICDCS 2016, Nara, Japan, June 27–30, 2016, pp. 344–353 (2016)

27. Wang, Y., Cai, Z., Tong, X., Gao, Y., Yin, G.: Truthful incentive mechanism with location privacy-preserving for mobile crowdsourcing systems. Comput. Netw. **135**, 32–43 (2018)

28. Xu, G., Li, H., Tan, C., Liu, D., Dai, Y., Yang, K.: Achieving efficient and privacy-preserving truth discovery in crowd sensing systems. Comput. Secur. **69**, 114–126 (2017)

29. Zheng, Y., Duan, H., Yuan, X., Wang, C.: Privacy-aware and efficient mobile crowdsensing with truth discovery. IEEE Trans. Depend. Secure Comput. **PP**(99), 1–1 (2017)

30. Zheng, Y., Duan, H., Wang, C.: Learning the truth privately and confidently: encrypted confidence-aware truth discovery in mobile crowdsensing. IEEE Trans. Infor. Forensics Secur. **13**(10), 2475–2489 (2018)

31. Zhang, C., Zhu, L., Xu, C., Sharif, K., Du, X., Guizani, M.: LPTD: achieving lightweight and privacy-preserving truth discovery in CIoT. Fut. Gener. Comput. Syst. **90**, 175–184 (2019)

32. Yuan, J., Yu, S.: Efficient privacy-preserving biometric identification in cloud computing. In: INFOCOM, pp. 2652–2660 (2013)

33. Li, Y., Gao, J., Meng, C., Li, Q., Su, L., Zhao, B., Fan, W., Han, J.: A survey on truth discovery. SIGKDD Explor. **17**(2), 1–16 (2015)

34. Miao, C., Jiang, W., Su, L., Li, Y., Guo, S., Qin, Z., Xiao, H., Gao, J., Ren, K.: Cloud-enabled privacy-preserving truth discovery in crowd sensing systems. In: Song, J., Abdelzaher, T.F., Mascolo, C. (eds.) Proceedings of the 13th ACM Conference on Embedded Networked Sensor Systems, SenSys 2015, Seoul, South Korea, November 1–4, 2015, pp. 183–196. ACM (2015)

35. Zheng, Y., Duan, H., Yuan, X., Wang, C.: Privacy-aware and efficient mobile crowdsensing with truth discovery. IEEE Trans. Depend. Secur. Comput. **17**(1), 121–133 (2020)

36. Wong, W.K., Cheung, D.W.L., Kao, B., Mamoulis, N.: Secure kNN computation on encrypted databases. In: Çetintemel, U., Zdonik, S.B., Kossmann, D., Tatbul, N. (eds.) Proceedings of the ACM SIGMOD International Conference on Management of Data, SIGMOD 2009, Providence, Rhode Island, USA, June 29–July 2, 2009, pp. 139–152. ACM (2009)

37. Shu, J., Jia, X., Yang, K., Wang, H.: Privacy-preserving task recommendation services for crowdsourcing. IEEE Trans. Serv. Comput. **14**(1), 235–247 (2021)

38. Tang, J., Fu, S., Xu, M., Luo, Y., Huang, K.: Achieve privacy-preserving truth discovery in crowdsensing systems. In Zhu, W., Tao, D., Cheng, X., Cui, P., Rundensteiner, E.A., Carmel, D., He, Q., Yu, J.X. (eds.) Proceedings of the 28th ACM International Conference on Information and Knowledge Management, CIKM 2019, Beijing, China, November 3–7, 2019, pp. 1301–1310. ACM (2019)
39. Zhang, C., Zhu, L., Xu, C., Liu, X., Sharif, K.: Reliable and privacy-preserving truth discovery for mobile crowdsensing systems. IEEE Trans. Depend. Secur. Comput. **18**(3), 1245–1260 (2021)
40. Miao, C., Su, L., Jiang, W., Li, Y., Tian, M.: A lightweight privacy-preserving truth discovery framework for mobile crowd sensing systems. In: 2017 IEEE Conference on Computer Communications, INFOCOM 2017, Atlanta, GA, USA, May 1–4, 2017, pp. 1–9. IEEE (2017)

# Part IV
# Summary and Future Research Directions

# Chapter 8
# Summary

**Abstract** In this chapter, we first summarize the research on privacy issues in MCS and the innovative points of the above chapters. Then, we provide the future research outlook for privacy issues in MCS.

## 8.1  Conclusion

Mobile crowdsensing (MCS), as a human-centered data sensing model, combines the ideas of mobile sensing and crowdsensing. It utilizes the mobile crowdsensing platform to connect sensing users and data requesters. In MCS, the mobile crowdsensing platform and users cooperate to complete many large-scale and complex social sensing activities, e.g., environmental monitoring, traffic management, and disease diagnosis. MCS provides great convenience for people's production and life and promotes the development of the Internet of Everything.

However, sensing users incur various privacy threats during participating in mobile crowdsensing applications. The leakage of private information will bring economic and property losses to users and even seriously threaten their life safety. Therefore, preserving user privacy plays an important role in mobile crowdsensing. In this book, we first analyze the privacy issues in mobile crowdsensing. Next, for the two phases of data collection and processing, we propose corresponding privacy-preserving solutions for different phases. The innovation points of this book are summarized as follows:

- In the data collection phase, we propose a privacy-preserving content-based task allocation scheme, named PPTA, in Chap. 3 to address the problems of the existing privacy-preserving scheme with weak security, single function, and poor task allocation efficiency. PPTA utilizes the polynomial function to aggregate task information, utilizes the randomizable matrix multiplication technique to effectively preserve task and query privacy, and utilizes the nature of the polynomial function and matrix to achieve efficient task allocation. With strong generality and practicality, PPTA can support privacy-preserving conjunctive task allocation, threshold task allocation, task allocation with access control, task restoration, etc.

- In the data collection phase, we propose a privacy-preserving geometric range-based privacy-preserving task allocation scheme, named GPTA, in Chap. 4 to address the problems of weak security, restricted query range, and poor location retrieval efficiency of existing privacy-preserving schemes. GPTA utilizes the polynomial fitting technique to efficiently and effectively fit arbitrary geographic ranges, the random matrix multiplication technique to efficiently preserve location and query privacy, and the nature of the matrix for efficient location retrieval. The dual-server model in GPTA can effectively resist collusive attacks by the single server and users. In addition, GPTA constructs a query index structure from the query history of the data requester to achieve nonlinear location retrieval efficiency.

- In the data processing phase, we propose a privacy-preserving truth discovery scheme with truth transparency, named RPTD, in Chap. 5 to address the problems of weak privacy-preserving and low computational and communication efficiency of existing privacy-preserving schemes. Specifically, RPTD-I is designed for scenarios where users can participate in the truth discovery iterative process, which achieves efficient weight and truth updates with the privacy of data and weights. Subsequently, RPTD-II is designed for scenarios where users cannot participate in the truth discovery iteration process. By transferring the computational operations of users in the truth discovery iteration to the cloud server, RPTD-II greatly reduces the computational and communication overhead on the user side.

- In the phase of data processing, we propose a privacy-protecting truth discovery scheme (SATE) with truth hiding in Chap. 6, which improves the efficiency of computation and communication. SATE utilizes two non-colluding cloud servers to construct the mobile crowdsensing platform and utilizes data perturbation technology and the public-key cryptosystem with cooperative decryption to achieve weight and truth updates without compromising the privacy of data, weights, and truth.

- In the data processing phase, we propose a privacy-preserving truth discovery scheme with task hiding (LPTD) in Chap. 7, which improves the computational efficiency and communication efficiency. LPTD utilizes the techniques of secure k-nearest neighbor, data perturbation, and matrix decomposition to preserve data privacy and task privacy and achieves weight and truth updates without compromising the privacy of data and task.

## 8.2  Outlook

- **Nonlinear task allocation efficiency.** Most of the existing privacy-preserving task allocation schemes can only achieve task allocation efficiency with linear time complexity. With the increase in task number, large computational overhead will be introduced on the mobile crowdsensing platform. Most existing solutions with nonlinear retrieval efficiency are based on symmetric encryption algorithms and rely on a trusted third party to build data index structures in plaintext, which is difficult to implement in mobile crowdsensing. We may consider the historical queries of data requesters and build a data index structure in ciphertext form, aiming to achieve more efficient task allocations without compromising user privacy.
- **Reputation value-based task allocation.** Reputation-based task allocation mechanism is also an important component of mobile crowdsensing. Most of the existing reputation-based task allocation schemes only consider reputation value comparison, which however ignore the generation, management, and verification of reputation values. We may consider designing privacy-preserving reputation value management schemes that can simultaneously support reputation calculation, verification, and prediction, without disclosing the privacy of user reputation values.
- **Blockchain-based truth discovery and reward distribution.** Blockchain has received increasing attention for its open and decentralized properties. The participation of sensing users in mobile crowdsensing usually results in monetary rewards. The fair distribution of rewards based on users' data quality is an important research component of mobile crowdsensing. We may consider designing a blockchain-based incentive mechanism, using truth discovery to calculate user weights and using blockchain and smart contracts to guarantee user benefits to motivate users to actively participate in mobile crowdsensing applications and provide high-quality data.
- **Privacy-preserving machine learning.** Machine learning algorithms can mine valuable models from user-submitted data and are widely used in scenarios such as environmental monitoring, smart healthcare, and smart factories. Existing privacy-preserving machine learning schemes incur large computational and communication overheads on the user side. We may consider designing efficient privacy-preserving machine learning schemes by using data perturbation techniques, random matrix multiplication techniques, secure multi-party computation, etc.

Printed in the United States
by Baker & Taylor Publisher Services